DATE DUE

D1074563

8/27/82

Annals of the CEREMADE

Edited by
J. P. Aubin
A. Bensoussan
I. Ekeland

Birkhäuser
Boston · Basel · Stuttgart

Mathematical Techniques of Optimization, Control and Decision

J. P. Aubin,
A. Bensoussan,
I. Ekeland,
editors

1981

Birkhäuser
Boston • Basel • Stuttgart

Editors:

J. P. Aubin
C.E.R.E.M.A.D.E.
Université de Paris IX Dauphine
F-75775, Paris Cedex 16
FRANCE

A. Bensoussan
C.E.R.E.M.A.D.E.
Université de Paris IX Dauphine
F-75775, Paris Cedex 16
FRANCE

I. Ekeland
C.E.R.E.M.A.D.E.
Université de Paris IX Dauphine
F-75775, Paris Cedex 16
FRANCE

Library of Congress Cataloging in Publication Data

Main entry under title:

Mathematical techniques of optimization, control, and decision
(Annals of the C.E.R.E.M.A.D.E.)
Bibliography: p.
Includes index.
1. Mathematical optimization--Addresses, essays, lectures. 2. Control theory--Addresses, essays, lectures. 3. Decision-making.
I. Aubin, Jean Pierre. Bensoussan, Alain.
III. Ekeland, I. (Ivar), 1944- .
IV. Series.
QA402.5.M36 519 81-38452
ISBN 3-7643-3032-5 AACR2

CIP - Kurztitelaufnahme der Deutschen Bibliothek

Mathematical techniques of optimization, control, and decision
Ed. by J. P. Aubin
Boston, Basel, Stuttgart: Birkhäuser, 1981
(Annals of the C.E.R.E.M.A.D.E.)
ISBN 3-7643-3032-5
NE: Aubin, Jean-Pierre [HRSG.]

All rights reserved. No part of this publication may be reproduced, stored in a retrieval system, or transmitted, in any form or by any means, electronic, mechanical, photocopying, recording or otherwise, without prior permission of the copyright owner.

© Birkhäuser Boston, 1981

ISBN: 3-7643-3032-5

Printed in USA

VANDERBILT UNIVERSITY
LIBRARY
NASHVILLE, TENNESSEE

CONTENTS

PREFACE

This book is intended to be the first of a series. Its purpose is to provide reports on non-standard aspects of applied mathematics, centered mainly around decision making. The mathematical problems one is led to are quite varied, and so are the techniques used to solve them. This is why the contributions in the volume cover so broad a range, from ordinary to partial differential equations, from Banach space theory to game theory. But under this variety of method lies a unity of purpose, as J. P. Aubin explains in the general overview which opens this volume.

All the contributions were initiated as colloquium talks at the C.E.R.E.M.A.D.E. during the 1979 - 1980 academic year. We thank the contributors for making their work available for this volume.

J. P. AUBIN A. BENSOUSSAN I. EKELAND

CONTRIBUTORS

J. B. AUBIN, C.E.R.E.M.A.D.E., Université de Paris IX Dauphine, F-75775, Paris Cedex 16, FRANCE

A. BENSOUSSAN, C.E.R.E.M.A.D.E., Université de Paris IX Dauphine, F-75775, Paris Cedex 16, FRANCE

P. BERNHARD, C.E.R.E.M.A.D.E., Université de Paris IX Dauphine, F-75775, Paris Cedex 16, FRANCE

J. FREHSE, Institut für Angewandte Mathematik, Universität Bonn, Beringstr. 4-6, 5300 Bonn 1, West GERMANY; and C.E.R.E.M.A.D.E., Université de Paris IX Dauphine

N. GHOUSSOUB, Department of Mathematics, University of British Columbia, V6T 1W5 Vancouver B.C., CANADA; and C.E.R.E.M.A.D.E., Université de Paris IX Dauphine

H. JONGEN, Universität Hamburg, Institut für Angewandte Mathematik, D-2 Hamburg 13, Bundesstrasse 55, West GERMANY

G. LAFFOND, C.E.R.E.M.A.D.E., Université de Paris IX Dauphine, F-75775, Paris Cedex 16, FRANCE

H. MOULIN, C.E.R.E.M.A.D.E., Université de Paris IX Dauphine, F-75775, Paris Cedex 16, FRANCE

J. ORTMANS, C.E.R.E.M.A.D.E., Université de Paris IX, Dauphine, F-75775, Paris Cedex 16, FRANCE

G. TROIANIELLO, Instituto Matematico Guido Castebruovo, Universita di Roma, Rome, ITALY; and C.E.R.E.M.A.D.E., Université de Paris IX Dauphine

G E N E R A L   O V E R V I E W

GENERAL OVERVIEW

J. P. Aubin

C.E.R.E.M.A.D.E.
Université de Paris IX Dauphine

C.E.R.E.M.A.D.E. stands for "Centre de Recherches de Mathématiques de la Décision"--Research Center in the Mathematics of Decision Making. From its foundation in 1971, the C.E.R.E.M.A.D.E. has drawn most of its problems from the *organization sciences*, management, economics, biology, systems theory, and this is where the techniques developed at the C.E.R.E.M.A.D.E. find their most immediate applications. The study of large systems, and the mathematics of decision making, are recurrent features throughout this research.

To be sure, this area of research has acquired widespread recognition and importance since the new industrial revolution. But it can also claim great antiquity: Systems theory was born in the 17th century to cope with the problems posed by mechanics and physics. Indeed, at the same time, its connection with optimization already appeared, albeit in a mysterious way (Fermat's, later Maupertuis', principle of least action). This is also the time when Euler laid the foundations of the Calculus of Variations.

The fact that mathematical concepts and techniques developed to solve problems in 18th century mechanics and physics have acquired a great importance in contemporary economics and management is a great comfort to mathematicians, always sensitive to the structuring and unifying power of the instrument they yield. The Lagrange multipliers, for instance, which first appeared in mechanics, have long since travelled to many other fields of science.

On the other hand, the fact that these problems have attracted for so long the attention of so distinguished mathematicians is quite chastening. In modern times, it has taken nearly three quarters of a century before the economic problems raised by Cournot, Edgeworth, Pareto, Walras, were translated into mathematics and subsequently solved.

All this may help to understand why the mathematicians at C.E.R.E.M.A.D.E. have tried to look for motivations and applications in mechanics and engineering as well as in the organization sciences. Comparison between these problems is possible and fruitful, providing structuring, simplification, in short, a better understanding of both. The C.E.R.E.M.A.D.E. encourages scientists with different backgrounds to share ideas and problems, and to meet together the challenge of modern societies.

Now that the aim is clear, we will give some indication on the work that has been done. After an introductory chapter, to show the kind of problems which arise in organization sciences, we classify subsequent material under the two themes of systems theory and decision making, the final chapter showing the interplay between them.

CONTENTS

I. MATHEMATICS OF DECISION MAKING

1. The complexity of organization sciences.

Organization sciences, for instance economics and management, face us with very complex mathematical problems. One could go so far as to say that these problems are more complex than those arising in physical sciences. The range of phenomena encountered in the physical sciences can be formalized by means of a limited number of equations. Moreover, with a suitable interpretation of variables and unknowns, these equations remain valid for very different phenomena. In many situations from physics or mechanics, sufficiently representative models of the fundamental phenomena are realized with a restricted number of equations. Each of these equations has given rise to a whole body of literature, for certainly the physical sciences are not simple. Nevertheless we claim that they are less complex than the organization sciences.

This is a miracle of nature which astonished Einstein and which has no reason to be universal. Consider, for example, instead of the organization sciences, biology, which cannot by any means be dealt with by a limited number of equations. We have very few ideas about any mathematical laws which govern living organisms.

We think that the situation of organization sciences falls between that of the physical sciences and those of biology.

To support this argument let us consider two examples. First that of a market economy where one finds very numerous decision makers, with vastly different size and behavior. They reach decisions in real time, with uncertain knowledge of the future, thus entering into a complex interplay of alliances and conflicts. What is the outcome of this game? Is it possible to determine a macroeconomic behavior that can be predicted and perhaps controlled?

Next consider an example in management, that of a large multinational firm which decides its investments, capital, production, advertisement, personnel management, policy with respect to its competitors, etc. Is it possible to optimize these decisions?

In these two cases we face complex phenomena which allow no simple model.

2. Possibility and use of modelling.

One could say, of course, and there are those who do, that this very complexity implies that mathematics is powerless to account for these phenomena and that other techniques must be used. And as we have mentioned biology, one thinks naturally of experimental, clinical and empirical methods.

These are obviously very useful, just as they are in the physical sciences, where they in no way obviate the usefulness of quantitative or theoretical methods. There is no conflict between experimental and theoretical physicists. The place of each is recognized.

In biology, on the other hand, the experimental approach is largely predominant. Our claim is that the situation in the organization sciences is--or ought to be--more akin to the one in physics.

Take the example of a multinational firm. At each instant it can be represented by a certain number of variables: The quantities produced, the stocks, the supplies coming in; The loans, the short, medium and long term investments, the capital in various currencies, the demand with its risks, advertisement assets, marketing factors; The situation of the capital, work force, investments. There are also the state of the environment, the existence of competitors, of subcontractors, cartels, etc.

It is not at all impossible to imagine and even to formulate relations between these variables, which let us recall, are functions of time, involving uncertainties of various kinds. Starting from this analysis, decisions can be made in such a way as to optimize the given criteria. There may be several of these, even for a single decision maker, and they usually vary with time. Finally, let us not forget that each decision maker will take into account what he believes will be the behavior of others and the general trend.

What renders the situation complex is less the nature of the relations than their sheer number and, of course, the uncertainty and imprecision of the data. Despite all this, organizations and especially firms must all solve this kind of problem, and others which shall be presented in the following pages.

This initial observation explains why we feel that mathematical techniques in the organization sciences or as we call it, the mathematics of decision making, must be developed together with research of a more empirical nature.

3. Mathematical economics.

Let us begin with economics. This is first of all a science, which implies that the economist seeks to put forward fundamental laws without immediately concerning himself about action. It was, therefore, inevitable that mathematical economics, as developed under the impetus of Walras, Pareto, Arrow, Debreu, Scarf, may sometimes seem extremely abstract and far removed from applications, as was the case when mathematical physics was developed.

Purely mathematical models have often predicted the existence of planets or certain elementary particles long before empirical evidence could be reached. In a similar way, models of mathematical economics which are developed abstractly, enrich our knowledge and may be useful in the future. The notion of market equilibrium price defined years ago is beginning to be used in concrete computations, thanks to recent algorithms, particularly for agricultural products. I.I.A.S.A. (International Institute for Applied Systems Analysis) has set up a whole program of research in this area, which is becoming extremely useful for the planning divisions of countries or organizations such as the E.E.C. (European Economic Community), the O.E.C.D. (Organization for Economic Cooperation and Development), etc.

The development of mathematical economics has required fundamental research in mathematics which goes far beyond the techniques and methods used in physics. New mathematical theories have been created for that purpose.

But fundamental research in mathematics is by no means restricted to mathematical economics. Econometrics, large macroeconomic models and maquettes have required the development of completely original algorithms. A new and important phenomenon appears here--the conjunction of mathematics of decision making and computer science, with an aim to immediate applications.

4. Management systems.

As for management, it must be concerned with concrete problems while economics are allowed a more abstract outlook. I would liken economics to physics or biology, and management to engineering or medicine. It is essentially an applied science, but the problems which it raises are not simple. As we have seen, one encounters complex dynamical systems which are little understood. The problem is to make decisions in such a way as to satisfy or optimize several criteria, often within situations of conflict. The more we try to reflect reality, the more we feel the need to develop new approaches, more and more sophisticated, which therefore require an enormous effort of fundamental and applied research.

5. The need for theory.

Control theory, first deterministic, then stochastic, and finally adaptative, has laid a firm claim on this area of applications. Indeed, management systems can be considered deterministic, stochastic, or adaptive, according to the level of sophistication which is required. The ultimate aim is to manage optimality and in real time inventories or financial assets. There is another very important field of applications for control theory, namely engineering. There is constant interaction between these two areas, and ever more frequently problems occur with technological and economic sides, closely linked.

Game theory also is essential to the organization sciences. Let us mention, among more recent developments, differential games, leader-follower games, voting procedures, multi-criteria optimization, the theory of teams, etc. All are strongly motivated by management: Advertisement, pricing, expansion, all the main policies of a firm lead to such situations.

There is no need to elaborate on the need for research in the mathematics of forecasting, model identification, sensitivity analysis, decomposition methods, hierarchical analysis. All this research is both fundamental and applied. It is applied because it attempts to arrive rapidly at implementable algorithms and even computer software. It is fundamental because it leads to complex theoretical problems.

Nevertheless in the case of management, in contrast to mathematical economics, one does not attempt to axiomatize a phenomenon in order to extract its structure. Theory can be justified only insofar as it is necessary for solving concrete problems. The motivation must ever remain present. This is why research is directed toward data processing and algorithms. The degree of difficulty remains exactly the same.

6. The limits of mathematical techniques.

Certainly one may object that all this effort is of interest only if in the end we dispose of data to feed into the models. This is correct, but here again progress is very rapid, thanks to computer science and more particularly to the development of mini computers, very well adapted to data preparation, and with prices which make them accessible even for medium-sized firms. Progress is also due to highly mathematized techniques of data analysis.

To be sure, we do not claim, that management and economics will use in the near or distant future only mathematical or computer techniques. What is true is that the combination of these techniques, together with data banks, will allow us to deal with a few important but specific problems in the life of organizations. The leadership

GENERAL PROBLEMS OF SYSTEMS THEORY

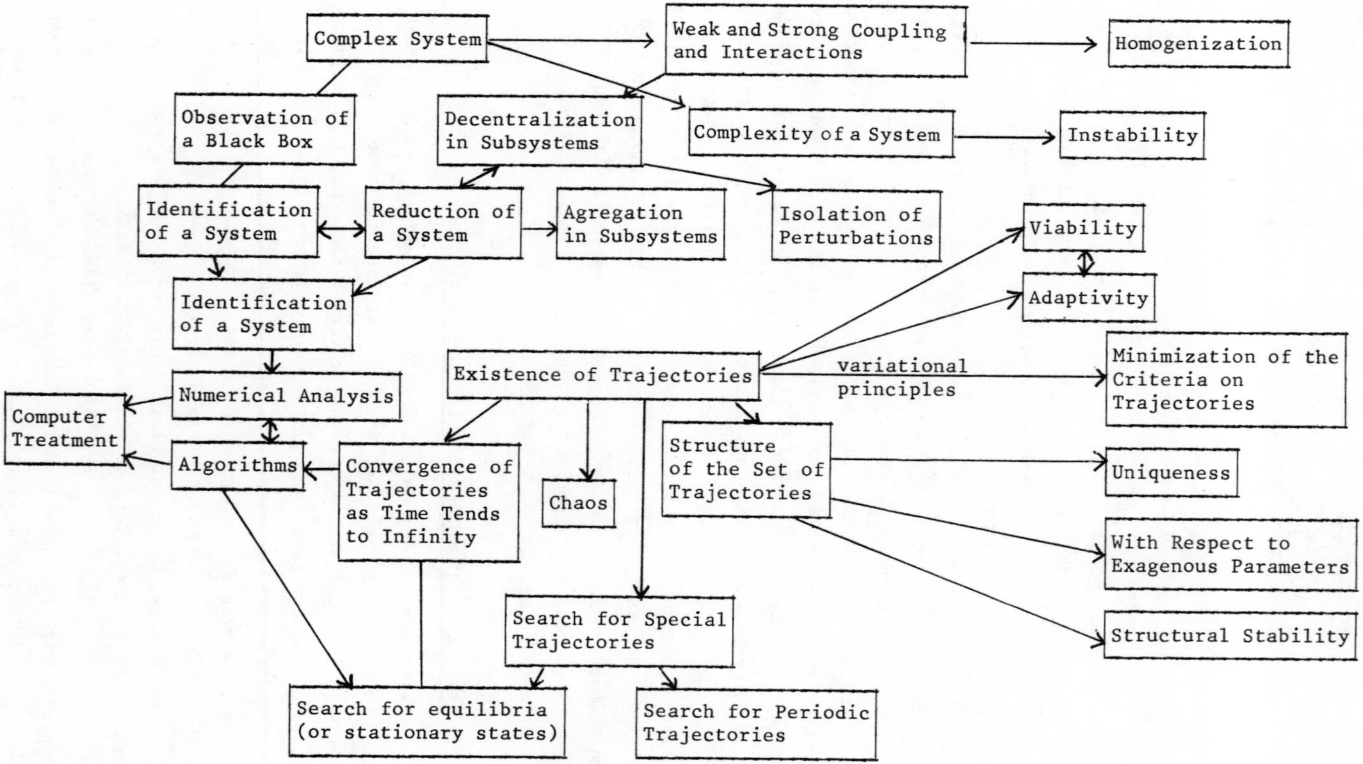
Complex System
Weak and Strong Coupling and Interactions
Homogenization
Observation of a Black Box
Decentralization in Subsystems
Complexity of a System
Instability
Identification of a System
Reduction of a System
Agregation in Subsystems
Isolation of Perturbations
Viability
Adaptivity
Identification of a System
Existence of Trajectories
variational principles
Minimization of the Criteria on Trajectories
Numerical Analysis
Computer Treatment
Algorithms
Convergence of Trajectories as Time Tends to Infinity
Chaos
Structure of the Set of Trajectories
Uniqueness
With Respect to Exagenous Parameters
Structural Stability
Search for Special Trajectories
Search for equilibria (or stationary states)
Search for Periodic Trajectories

problems, the legal, human and social problems, remain for the moment and perhaps for a long time beyond the reach of mathematical techniques. And even for sectors which make great use of quantitative techniques such as production, finance, marketing, human behavior certainly remains fundamental. What we should aim for is a coordinated use of all these disciplines in the areas where their efficiency is greatest.

II. SYSTEMS

1. Deterministic systems, stochastic systems and macrosystems.

A system is characterized by the evolution of its state as a function of time. This state evolves according to its own laws which we shall classify under the following three headings:

A. *Deterministic Systems:* The evolution law is deterministic. The velocity is a single-valued function of the state (past or present). In other words, the future states are determined by the initial state in a single valued fashion. There are models where one chooses to ignore the effects of perturbations or to represent them in a non-random way. They generally arise in mechanics, physics, engineering and any model which is voluntaristic (or finalistic, teleological or subject to economic planning) from the social sciences. The study of differential equations and difference equations which formalize these systems is at present very advanced.

B. *Stochastic Systems:* The evaluation law is called stochastic when the system is submitted to random perturbations (called noise in shoptalk). These systems mainly arise in engineering. They have obvious applications in numerous systems in management when the random component is sufficiently understood. Considerable progress has been made in the past twenty years in the study of these stochastic differential systems in which members of C.E.R.E.M.A.D.E. have actively participated.

C. *Macrosystems:* The evaluation law takes *uncertainty* into account in a crucial way (were this only because of the impossibility of an exhaustive description of the problem). We encounter these systems in macroeconomics, ecology or biology. They are also characterized by the *absence of controls* (or, what amounts to the same thing, by ignorance of the laws which relate the state of the system to the controls) and by *irreversibility* (in the sense that it is impossible to identically reproduce experiments). Naturally *these systems have no forecasting capacity* since it is useless to attempt to know exactly a system in which uncertainty plays such a role and experimentation is excluded.

The fruitful mathematical framework with which to approach the study of macrosystems is that of differential inclusions where the speed of evolution is a set-valued function of the state. In other words the initial state of such a system at each instant determines a more or less large set of possible states. The members of C.E.R.E.M.A.D.E. have made an essential contribution to this theory during the past few years.

2. Distributed systems and/or hereditary systems.

In fact the divisions are not as sharp as this classification seems to imply. For example, the theory of macrosystems yields very important information for determinist control theory. Similarly determinist differential games provide a possible model for control in the presence of (stochastic) perturbations, and these two theories have numerous interrelations.

Conversely, within a given class of systems the problems encountered can be profoundly different according to whether the state of the system has finite or infinite dimension. The latter case includes the case when the state is a function of other variables than time, for example, space variables. These are the systems which are called *distributed systems* and which are represented by partial differential equations.

In addition the C.E.R.E.M.A.D.E. has also been studying more complex problems which take into account past history. This is the case for instance when a system has its own response time. We call a system *hereditary* if the velocity of evolution depends not only on the state at a given moment but on all or part of the trajectory until the given moment (see the works of Haddad).

3. Existence problems, stability problems and their numerical analysis.

The mathematical framework is set forth: Differential equations, stochastic differential equations, differential inclusions. What are the problems?

First of all these are the classical mathematical problems on the *existence* of trajectories, on *uniqueness* or, if not, on the *structure of the set trajectories*.

Similarly problems of *stability* are systematically taken into consideration. This term has many uses, leading to many different problems. For instance, when the system depends on exogenous parameters how do the trajectories depend on these parameters? Is the dependence regular? There is a whole range of ways to define regularity in the sense that small variations of the exogenous parameters induce small variations of the trajectory. This leads us to *sensitivity analysis*. One can also investigate whether a given property of a system it still enjoyed by all, or almost all, neighboring systems. This is the problem of structural stability.

Next comes the problem of *numerical analysis* of these systems: Can we transform and approach the problem in such a way as to render it suitable for treatment by computers and to obtain numerical values for the trajectories? We have already mentioned this question and we shall come back to it. Let us conclude by another question the C.E.R.E.M.A.D.E. has made important contributions to: How do trajectories behave for very large time? The solution of this problem enables us to characterize optimal paths in infinite horizon economic models.

4. The viability problem.

A new class of problems has arisen in recent years. They deal with the *viability* of trajectories, i.e. the possibility for a given evolutionary system to satisfy constraints imposed by the environment, for instance scarcity constraints in economics (the excess demand must be negative or zero) or in ecology (the sum of the resources consumed is limited). This term is synonymous with *homeostasis*, which was coined by the biologist Cannon in 1932. The problem of viability was taken up by Claude Henry in 1972 with the framework of *planning methods* and is the source of important developments to which C.E.R.E.M.A.D.E. has actively contributed through work of Aubin, Clarke, Cornet and Haddad. The problem of isoperimetric constraints in the calculus of variations also falls within this framework.

5. Control Systems: Open loop and closed loop controls.

Using the mathematical framework of differential inclusions, we can recast the problems of control theory. This is the study of dynamical systems where the velocity at any instant depends, not only on past history and/or present state of the system, but also on a suitable control to be optimally chosen in a prescribed set.

An open-loop control depends only on time. A closed-loop control, or feedback, depends also on the state of the system. This offers the advantage of automatically correcting the trajectory, should the system steer away from its optimal path.

On the other hand, closed-loop controls are much harder to come by. The problem of finding feedbacks for which the trajectories of a system will satisfy a set of properties given in advance is very difficult in general.

For example, one could ask for viable trajectories in applications to economics, or for stable trajectories, in applications to systems theory.

In section IV we shall come back to the fundamental properties played by feedbacks in optimal control theory and differential games (determinist and stochastic).

6. Adaptative Systems: The Darwinian approach.

We shall now consider a whole series of problems relative to the selection of trajectories. In the absence of uniqueness, either because of uncertainty or because of the precense of explicit controls, the problem arises of finding methods for selecting trajectories.

We can classify these methods under two headings:

A. Problems where one or several decision makers intervene capable of making forecasts about the future and able to direct the system by means of controls, making their decisions once and for all at the initial instant (even if these decisions depend on future events of which only the probability is known).

These methods which consist in optimizing the criteria (with the possibility of conflict) belong to optimal control theory and the theory of differential games to which Section IV of this report is devoted.

B. Problems where the system itself selects trajectories which improve its state as time elapses. This requires that the set of possible states be preordered, in the sense that some states are considered preferable to the current one. This occurs for instance in economics, where the behavior of agents is submitted to an utility function which they strive to increase as time elapses. This allows us to take into account H. Simon's concept of "satisfying" (rather than "optimizing").

These are called Darwinian, or adaptative systems, according to the area of applications one has in mind. Typically, they lead to two kinds of mathematical investigations:

A. Knowing the system, to find which are the compatible preorderings.

B. Knowing the preording, to construct the dynamical system.

This second approach, which is systematically treated in the work of Cornet, turns out to be very fruitful. It allows one to study the procedures of economic planning (of the type proposed by Malinvaud, Drèze, de la Vallée Poussin), to discover others, to classify them and to compare their respective properties.

The work of Champsaur, Cornet, Henry, etc. on this subject has shown that certain planning procedures lead to what are called *variational differential equations* introduced on the basis of problems arising in continuum mechanics and optimal control theory. See the book by Duvaut and Lions, "Inéquations en Mécanique et en Physique," that of Lions, "Sur quelques questions d'analyse, de mécanique et de contrôle optimal" and the work of Attouch, Brézis, Cornet, Damlamian, etc.

We recover as well, with considerable improvement, the theory of stability in the sense of Liapounov (the bases of which were formulated in the last century) which concerns the behavior of the trajectories of a system as time tends to infinity. Note also that the dynamical programming of Bellman, (which goes back at least to Carathéodory), falls into the framework of this theory (see the last part of this report).

Such comparisons have brought to these diverse problems structuring, simplification, greater clarity and in each case, a deeper understanding.

7. Complexity, chaos, and catastrophes.

In many different areas of science, there has been a renewal of interest in the dynamics of very large deterministic (but uncontrolled) systems. The very notions of complexity and stability are subject to investigations, as in R. May's work on ecosystems. There are several mathematical approaches.

The first one deals with systems which have a large number of internal variables, but depend on few (at most six) parameters, René Thom has given conditions under which there exists frontiers, the crossing of which by the parameters leads to discontinuous response of the systems--jumps, or more poetically, catastrophes. The general shape of these frontiers is well understood.

The second one looks for erratic behavior. It is often the case that even relatively small systems, such as those of celestial mechanics, although completely deterministic, closely mimic stochastic systems. In other words although each individual trajectory is completely predictable, the global picture is almost completely unpredictable. For instance, tossing a coin is universally considered a game of chance, although in principle it is a game of skill. The outcome is completely determined by the way the coin is thrown (initial condition); the only problem is that no one has skill enough to throw the coin so finely that it will land precisely on heads.

The third one analyses complexity by successive "bifurcations" starting from a simple situation. This is how Ruelle and Takens study turbulence. This point of view begins to be used in dynamical economics.

8. Decomposition and coupling of large systems.

The decomposition of a complex into subsystems provides a whole source of problems. Another area of interest is the properties of couplings which guarantee that the stability of certain subsystems implies the stability of the whole system.

The problem in which there are weak and strong interactions between different parts of a system come within the theory of *singular perturbations* and of *homogenization*.

9. Identification of systems.

We now take up the problems of identification. We consider a system whose structure is known exactly and which is a function of unknown parameters. We observe its trajectories and seek to identify the model, that is to adjust the unknown parameters in such a way that the model restitutes the available observations (this problem is often called the *inverse problem*).

This problem is familiar in econometrics where one of the objectives is to identify finite dimensional systems. The identification of infinite-dimensional systems has fundamental applications in the domain of energy. Methods elaborated by Chavent (who studied not only the theoretical aspect, but also numerical techniques for computing solutions) have been used in a model of a natural gas field in Holland managed by Elf-Aquitaine. Another model used by the same firm proposes an algorithm which allows real seismic recordings to be dealt with.

10. The black box: Internal representation of systems.

We arrive now at the central problems of systems theory. The most famous is that of the black box, or of the internal representation of a system: We know the law which associates to every input of the system its output, but we do not know the states of the system, not its dynamics. We seek a space of states (as small as possible) and differential equations which model the observed input relationships as closely as possible. The techniques used by the vast majority of specialists are algebraic. If they are well adapted to describe discrete systems, they are less so for systems where time varies continuously: Aubin and Bensoussan have proposed other techniques which solve this problem.

11. Infinite horizon, equilibria and algorithms.

In certain situations, which we set ourselves to discover, it is shown that as time tends to infinity, the trajectory converges to a constant trajectory which is called the *equilibrium* (or *stationary state*) of the system.

Let us quote the example of the Walras tâtonnement, an evolution system (the market price evolves as a function of the excess demand) which we wish to show converges to a constant price which is the Walras price equilibrium. Despite the progress made in this domain, this problem and analogous ones remain open.

Thus we take up two series of problems:

A. To show the existence of the properties of the equilibria of a given system.

B. To show under what conditions the trajectories converge to equilibria.

The connection between systems theory and nonlinear analysis enables one to prove that under certain conditions *the existence of viable trajectories from each initial state implies the existence of an equilibrium*. The convergence of trajectories to (economic) equilibria or Pareto optima is one of the objectives of the planning methods of which we spoke above.

There is also a numerical side to these questions. Indeed, one of the most efficient methods devised for computing equilibria and fixed points, Newton's method, is dynamic in character. We refer to the work of Lasry and Siconolfi in this volume.

12. Particular trajectories: Trajectories and equilibria of flux.

As we have seen, looking for an equilibrium is looking for a particular evolution: In this case, a constant trajectory, that is an absence of evolution. We can also ask if there exist other types of evolution. A problem which has been the object of intensive research for a long time is that of seeking *periodic* trajectories. On this subject Ekeland and Lasry have just made an important contribution using techniques developed within the framework of the theory of optimization.

Another particular law of evolution which has a special importance in economics is that of the uniform rectilinear movement of our student days, but in several dimensions. In this case we have an equilibrium of flux (also called a dynamical equilibrium). Here it is the speed of evolution (not the state) which is constant. A necessary condition is that the law of evolution be a function of time: Thus we say, in the language of systems theory, that the system is open. In the framework of control systems, this brings us to the problem of looking for controls (feedback) which allow such trajectories.

The study of solitary waves and solutions falls in this category. Introduced in the framework of the Korteweg--De Vries equations, solitary waves can be used for explaining the evolutions of spokes along a neuron. Lasry is studying this problem in collaboration with biologists.

III. DECISION MAKING

1. Mechanisms of decision making.

In the final analysis this problem amounts to constructing mechanisms for making the optimal choice of one or several elements of a given set (of decision variables). These mechanisms must reflect as well as possible the real procedures of decision making.

Although, of course, we are far from formalizing these mechanisms satisfactorily, the scientists at C.E.R.M.A.D.E. have patiently made their contributions to the working out of methods or concepts which bring us closer to the goal. For the moment most of the work concerns the static case, whereas time and uncertainty must play a crucial role. However, even in the static case, the problems are difficult, and the mathematical tools used are both recent and specific.

Let us remark right away that some important mathematical theories (convex analysis, nonlinear analysis, generalized gradients, contingent derivatives, etc.) have been constructed in part to help answer the questions raised. Two books of Ekeland (Théorie des Jeux et ses Applications Mathématiques and Eléments d'Economie Mathématique), that of Aubin (Mathematical Methods of Game and Economic Theory) survey the results.

2. The case of one decision maker: Optimization.

The first type of mechanisms is that of optimization where one decision maker chooses to minimize a test function.

The problems of existence, uniqueness and stability are well known. The duality method has been evolved within the elegant framework of convex analysis precisely to solve them.

Members of the C.E.R.E.M.A.D.E. have worked to extend these results to infinite dimensional spaces (using more general constraint qualification conditions) and have also developed the more elementary framework of quadratic programming, by means of which explicit solutions of minimization problems are obtained. In this way principles of decentralization and decomposition allow the explicit solution of complex problems.

Recent contributions of Aubin and Ekeland have treated the case where convexity is no longer a hypothesis. They have introduced the concept of lack of convexity and have used it to extend duality to problems with integral criteria and constraints. Aubin has characterized the classes of functions which share the properties which are useful in optimizing convex functions. In 1972 Ekeland proved an approximate variational principle which since its publication has had considerable success and numerous applications. Ekeland and Lebourg used this technique in particular to prove that almost all optimization problems have a unique solution which can be computed.

Aubin and Clarke have shown that the property of Lagrange multipliers which is useful to economists, namely their role in measuring the marginal cost of constraints, remains valid without convexity. With the introduction of the contingent derivative of correspondence we will be able to complete the sensitivity analysis both of the optimal value and of the set of optimal solutions.

3. Two-person games.

Moulin has done authoritative work on the theory of two-person games. He has devised ways to get around the classical paradoxes of game theory by a deeper understanding of the concept of strategy, and has begun a classification of games on the basis of information structure, dissuasion and for cooperation. In collaboration with Gabay he has constructed an algorithm for convergence to non-cooperative equilibria.

The minimax and saddle point theorems of von Neumann have been

been improved and simplified. By replacing strategies by decision rules, Lasry has shown a dual inequality to that of Ky Fan: These inequalities are more powerful and more general than those of minimax. They allow each player to use decision rules instead of choosing strategies. These decision rules which associate to each strategy of the adversary one or several strategies of the player notably enrich the structure of the game.

4. Cooperative games and fuzzy games.

The introduction of fuzzy games by Aubin has opened a new and unifying point of view in the theory of cooperative games. The idea is to enlarge the set of coalitions which can be formed by accepting that players participate partially in a coalition. In other words we introduce a rate of participation which can vary from 0 to 1. This type of generalized coalition is called a fuzzy coalition.

The structure enlarged in this way has been used to show that concepts as apparently different as that of the Shapley value and that of the core of a game are but special cases of the same concept (marginal gain when the players join the total coalition). Moreover, this same structure has been used to prove that the Walras equilibria of an economy coincide with the set of allocations accepted by all the fuzzy coalitions.

The concept of fuzzy game, suitably adapted, can also be used for other models. For example, by considering both positive and negative rates of participation, we can take into account different degrees of cooperation or aggressivity of the players in a coalition. In this way we can work out models of formation of coalitions and of their evolution, problems which need to be studied thoroughly. The same structure has been enriched by introducing the idea of a behavior profile, in order to use it in psycho-sociology.

This theory can also lead to defining power indices (size of the set of winning fuzzy coalitions in which a player participates). Roure has defined and used these indices to single out from the network of interownership among firms the groups and their leaders.

5. Mechanisms of decision making in economic and political theories.

Recently Moulin and his students have taken up again the study of the mechanisms of decision in economics and political theories. Positive political theory (that is the study of actual voting procedures) and the influence of information structures in economics (the problem of revealing preference and the study of tactical lies which every decentralized decision making mechanism induces in the individual agents) have known during the past five years a common formalization springing from game theory.

The generality of this approach has yielded a deeper understanding of the Walras model of general equilibrium by allowing a real strategic analysis of underlying micro-economic behaviors. This new theoretical point of view is full of promise which is limited only by the need for the conceptual refinements which it requires.

6. Regulation mechanisms: An approach to desequilibrium.

A new approach to desequilibrium in economics has been proposed by Aubin by taking into account time and evolution in the problem of allocation of scarce resources. For instance in the framework of an exchange economy, consumers must share at each instant an available commodity: Such a succession of choices will be called a viable trajectory.

It is assumed that a consumer is an automation able to associate with a price system and its past consumptions a rate of change in its present consumption. In other words, each consumer obeys a differential equation with memory depending on a common parameter, the price system. This describes a decentralized mechanisms regulated by a control (the price) summarizing the relevant information. By requiring budgetary rules (playing the role of Walras laws in the static case), one can prove the existence of prices depending upon the time such that the associated consumption trajectories remain viable.

Actually, these prices depend upon the time through the consumptions: They are closed-loop controls.

Other regulation mechanisms can be treated by this approach: Regulation by shortages or by introducing fiduciary goods.

DECISION MAKING

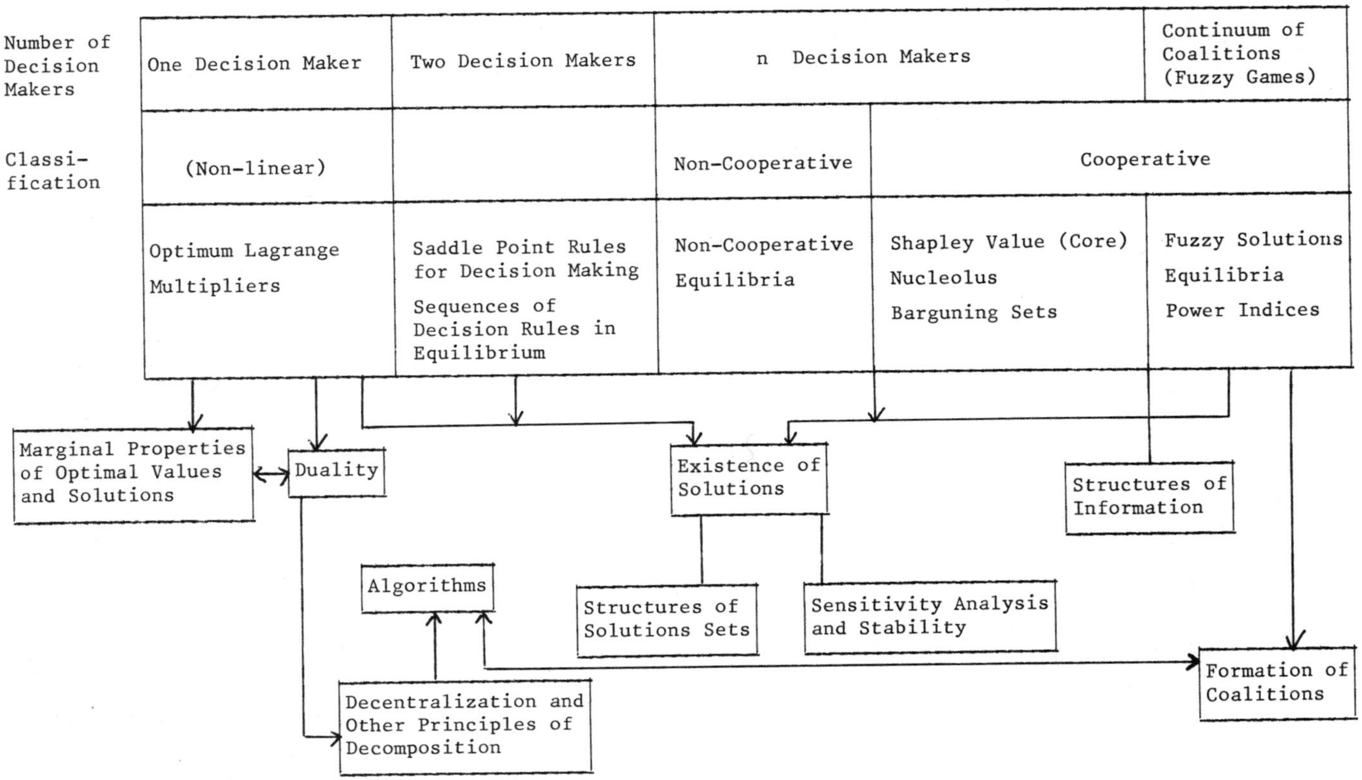
Number of Decision Makers
One Decision Maker
Two Decision Makers
n Decision Makers
Continuum of Coalitions (Fuzzy Games)
Classi-fication
(Non-linear)
Non-Cooperative
Cooperative
Optimum Lagrange
Multipliers
Saddle Point Rules for Decision Making
Sequences of Decision Rules in Equilibrium
Non-Cooperative
Equilibria
Shapley Value (Core)
Nucleolus
Barguning Sets
Fuzzy Solutions
Equilibria
Power Indices
Marginal Properties of Optimal Values and Solutions
Duality
Existence of Solutions
Structures of Information
Algorithms
Structures of Solutions Sets
Sensitivity Analysis and Stability
Decentralization and Other Principles of Decomposition
Formation of Coalitions

IV. SYSTEMS AND DECISION MAKING

1. How to select a trajectory.

Under this heading we group together the problems concerning the selection of trajectories of an evolution system, that is the study of the mechanisms of decision making when the decision variables are the trajectories of a system. These problems use the techniques described in the preceeding sections, and new ones as well.

Calculus of variations, optimal control theory, differential games fall in the domain of this kind of problem. Mechanics and physics had already provided numerous variational principles, postulating that nature minimized certain functionals on a family of virtual trajectories. Euler and Lagrange had already deduced the equations of movement from this. This work and that which followed from it constitutes what is called the calculus of variations to which optimal control theory has recently been added. Here again the problem is to minimize a functional on a set of trajectories. In the calculus of variations the functional takes into account the position and the velocity along the trajectory, and in optimal control theory it is a function of the state and of the control.

In both these cases optimization problems are obtained which are more difficult than those of mathematical programming (the decision variables which are functions of time run over infinite dimensional spaces rather than finite dimensional ones). These supplementary difficulties make these problems challenging; in any case we must tackle them if we want to represent the effects of time and variable behaviors in the systems we are studying.

All the members of C.E.R.E.M.A.D.E. have contributed to these areas as bear witness the books of Aubin (Mathematical Methods of Game and Economic Theory), of Bernhard (Commande Optimale, Décentralisation et Jeux Dynamiques), of Bensoussan, Hurst and Naslund (Management Applications of Modern Control Theory), of Bensoussan and Lions (Applications des Inéquations Variationnelles en Contrôle Stochastique and Contrôle Impulsionnel et Inéquations Quasi-variationelles) and of Ekeland and Teman (Analyse Convexe et Problèmes Variationnels).

SYSTEMS AND DECISION MAKING: SELECTION OF TRAJECTORIES

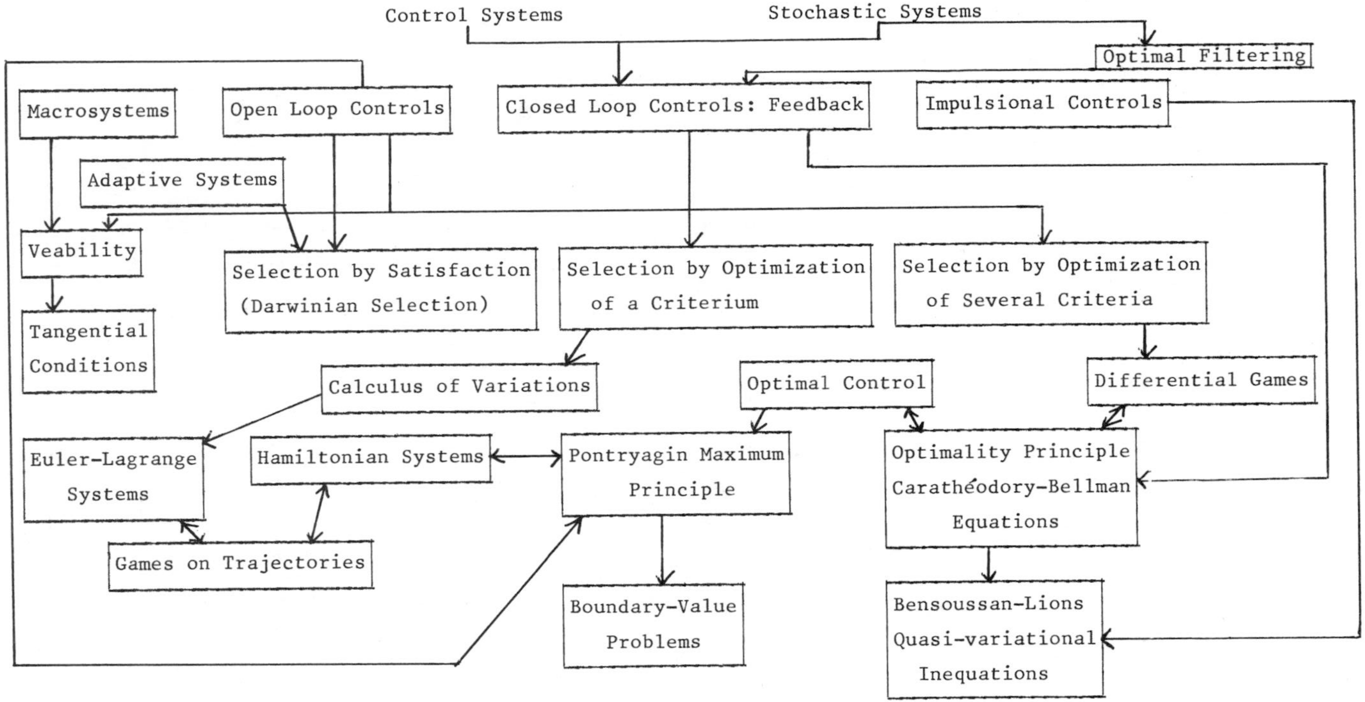

2. The maximum principle: Open loop controls.

Since Pontryagin, mathematicians have attempted to prove the maximum principle in various contexts and for different kinds of problems. The approach followed at C.E.R.E.M.A.D.E. has been to treat these problems as optimization problems on appropriate functional spaces. This point of view has not only allowed considerable simplification of classic proofs but also the successful solution of new problems. To cite only one example, Arrow devoted his Nobel prize lecture to the use of optimal control theory in infinite horizon and mentioned many paradoxes. The approach indicated above allowed Bensoussan, Aubin and Clarke to resolve them and allowed a whole school of mathematical economists to apply the maximum principle in infinite horizon. It has also been fruitful in the case where the convexity hypotheses are lacking and where the constraints concern the state of the system.

On the subject of convexity, Ekeland has shown what is called a relaxation property for replacing (in fact, approximating) a non-convex problem by a problem where convexity is present and which can, therefore, be treated by the methods of convex analysis.

3. Optimality principle: Closed loop controls.

Another point of view from which these problems can be treated goes back to Hamilton and Jacobi and has been explicitly developed by Carathéodory and widely used by Bellmann (optimality principle). It consists in constructing an optimal strategy step by step by going back in time. If the case of discrete systems does not present insuperable difficulties, this is not so for systems where time is a continuous variable.

It happens that this approach is particularly well adapted to the case of stochastic optimal control theory: In this case the Carathéodory-Bellmann equations become better structure partial differential equations which can be solved by techniques currently available (see the work of Bensoussan and Lasry). Hence we are able to construct optimal feedbacks. The deterministic case is much more delicate and requires the introduction of appropriate techniques.

4. Impulsive control theory and inventory management.

This approach can also be used to work on problems of optimal and impulsive control theory. The latter was formulated in 1972 by Bensoussan when he was trying to modelize problems of inventory management: The problem is to know when to renew the stocks and by how many units. He showed, in collaboration with Lions, that the Carathéodory-Bellman equations became what are called quasi-variational inequalities. Curiously these quasi-variational inequalities already existed in game theory (although in finite dimension). We can place in this framework the famous theorem of Arrow-Debreu (1954) from which could be proved (for the first time) the existence of a Walrus price equilibrium.

5. Optimal filtering.

The special case of linear systems where the criterium to be minimized is quadratic has close connections with the algebraic theory of optimal control theory. The problem consists in estimating the state of a dynamical system on the basis of incomplete observations with random perturbations (white noise). It often occurs in conjunction with a control problem. We refer to the survey paper of Bernhard in this volume, and to the book of Bensoussan "Filtrage Optimal des Systèmes Linéaires."

6. Differential games.

When we associate with dynamical systems game theory, we obtain the theory of differential games. This theory presents many difficulties of a technical nature which explains the fact that it has known rather divergent developments. Some consist in finding specific methods for solving given games, following the work of Isaacs and Breakwell. Others try to create a mathematical framework in which one can guarantee the existence of solutions without being able to calculate them. Bernhard has made progress in both directions, solving games on computers and creating a formal framework in which he justifies these solution methods, the principle of feedback, etc. He has also made connections with the theory of perturbed systems, the quadratic linear theory, etc.

7. Variational principles and Hamiltonian systems.

Let us come back to the maximum principle in the calculus of variations and in optimal control theory. This leads to differential systems of a special kind called Hamiltonian systems.

Historically these systems arose in mechanics and appeared in many domains of science. Recently an entire issue of the Journal of Economics Theory was devoted to economic applications of Hamiltonian systems.

These systems have been studied by a large number of mathematicians for centuries and continue to be studied. It turns out that considerable progress has just been made by Ekeland in collaboration with Aubin, Clarke and Lasry. In fact a constant concern of his has been to establish a connection between differential equations and optimization theory. Already in his book with Temam the characterization of a class of equations (called elliptic equations) by problems of minimization of trajectories was made in the framework of convex analysis and successfully used. In collaboration with Brézis, then with Aubin, he has proved that some important classes of evolution equations (parabolic, then hyperbolic) in fact come from variational problems which he has tackled by methods of duality.

The principal success has been obtained in collaboration with Clarke: They have succeeded in formulating a principle of least action dual to that of Maupertuis and technically much easier to handle. By this entirely new method a series of results of primary importance have been obtained. Since Poincaré, looking for periodic solutions has been one of the only means we dispose of for tackling Hamiltonian systems. But no global method was available. The least action principle of Clark-Ekeland provided one, and immediately many problems were solved which had been open since the work of Lyapounov one hundred years ago. We quote the most difficult, solved in collaboration with Lasry: A Hamiltonian system with n degrees of freedom possesses at least n periodic trajectories on each energy level.

C O N V E X   A N A L Y S I S

VARIATIONAL PRINCIPLES FOR DIFFERENTIAL EQUATIONS OF ELLIPTIC, PARABOLIC AND HYPERBOLIC TYPE

J. P. AUBIN

C.E.R.E.M.A.D.E.
Université de Paris IX Dauphine

INTRODUCTION

Let ϕ be a convex differentiable function defined on Hilbert space H. We consider the following three types of equations:

"Elliptic Equations"

(1) $$-\ddot{x} + \nabla\phi(x) = 0, \quad x(0) = x(T) = 0.$$

"Parabolic Equations"

(2) $$\dot{x} + \nabla\phi(x) = 0, \quad x(0) = x_o.$$

"Hyperbolic Equations"

(3) $$\ddot{x} + \nabla\phi(x) = 0, \quad x(0) = x(T) = 0.$$

Variational principles for elliptic equations have been known for a long time indeed: Elliptic Equations (1) are Euler-Lagrange equations for the minimization of the convex Lagrangian $\int_0^T(\phi(x) + (1/2)||\dot{x}||^2)dt$. For "Hyperbolic second-order equations" (3) [the Newton equation for instance], the Lagrangian becomes $\int_0^T(-\phi(x) + (1/2)||\dot{x}||^2)dt$ and is no longer convex, so that its minimization no longer plays an important role. But, for this class of problem, the Hamiltonian $H(x,p) = \int_0^T(\phi(x) + (1/2)\|p\|^2)dt$ is convex, and we will use this fact for characterizing solutions of (3) by a variational principle combining the principle of least-action and its dual. So, we shall distinguish among the problems of calculus of variations the ones associated with convex Lagrangians (elliptic) from the ones associated with convex Hamiltonians (hyperbolic).

We also shall recall the analogous variational principle of Brézis and Ekeland for parabolic equations. We give indications about how to use these variational principles for proving existence of solutions.

We adopt the following outline:

1. Some easy concepts of convex analysis
2. Elliptic equations
3. Parabolic equations
4. Hyperbolic equations
5. Solving the Newton equation
6. The principle of least action and its dual
7. Conclusions.

1. Some easy concepts of convex analysis

Let V be a Hilbert space and $\phi: V \to]-\infty,+\infty]$ be a function which is finite at one point at least. If $p \in V^* = \mathcal{L}(V,\mathbb{R})$, we set

$$(1) \qquad \phi^*(p) = \sup_{x \in X} [\langle p,x\rangle - \phi(x)]$$

This defines a convex lower semicontinuous function from V^* to $]-\infty,+\infty]$. The same procedure yields the biconjugate function $\phi^{**}: V \to]-\infty,+\infty]$. We shall use the basis fact: $\phi = \phi^{**}$ *if and only if* ϕ *is lower semicontinuous and convex.*

Since definition (1) implies the more symmetric form

$$(2) \qquad \forall x \in V, \quad \forall p \in V^*, \quad \langle p,x\rangle \leq \phi(x) + \phi^*(p),$$

we are led to distinguish the pairs $(x,p) \in V \times V^*$ such that

$$(3) \qquad \langle p,x\rangle = \phi(x) + \phi^*(p).$$

In this case, we set

$$(4) \qquad p \in \partial\phi(x) \quad \text{or} \quad x \in \partial\phi^*(p).$$

The closed convex subset $\partial\phi(x)$ *is called the subdifferential of* ϕ *at* x.

If ϕ is convex and differentiable at x, then $\partial\phi(x) = \{\nabla\phi(x)\}$. One can prove also that if ϕ is convex and continuous at x, then $\partial\phi(x)$ is non-empty and bounded. If ϕ is convex and lower semicontinuous, then $\partial\phi(x)$ is non-empty when x ranges over a dense subset of $\mathrm{Dom}\ \phi = \{x \in V \text{ such that } \phi(x) < +\infty\}$.

If ϕ is convex and differentiable, the minimum in $\phi^*(p)$ is achieved when $0 = \nabla(\langle p,\cdot\rangle - \phi(\cdot))$, i.e., when $p = \nabla\phi(x)$. So, if $\nabla\phi$ is invertible,

$$(5) \qquad \phi^*(p) = \langle p, \nabla\phi^{-1}(p)\rangle - \phi(\nabla\phi^{-1}(p)).$$

Hence the conjugate function coincides with the *Legendre transform*.

The subdifferential of a lower semicontinuous convex function carries many of the properties of the usual gradient.

We shall use the following property.

PROPOSITION 1. Let $\phi: U \times V \to]-\infty,+\infty]$ be convex lower semicontinuous and finite at one point at least. We set

$$(6) \qquad \psi(x,p) = \sup_{v \in V} [\langle p,v\rangle - \phi(x,v)].$$

The function ψ is concave with respect to x, convex with respect to p. Also, $(f,p) \in U^* \times V^*$ belongs to $\partial\phi(x,v)$ if and only if $f \in \partial_x(-\psi)(x,p)$ and $v \in \partial_p\psi(x,p)$.

Proof. Let $(f,p) \in \partial\phi(x,v)$. We write $\phi(x,v) - \phi(y,w) \le \langle f,x-y\rangle + \langle p,v-w\rangle$ for all $y \in U$, $w \in V$. By taking $y = x$, we find $p \in \partial_v\phi(x,v)$, i.e. $v \in \partial_p\psi(x,p)$ or $\psi(x,p) + \phi(x,v) = \langle p,v\rangle$. So, we can write

$$\langle p,w\rangle - \phi(y,w) \le \langle f,x-y\rangle + \psi(x,p).$$

By taking the supremum with respect to w, we deduce that

$f \in \partial_x(-\psi)(x,p)$. The converse statement is obvious.

EXAMPLE. Let Ω be an open subset of $\mathbb{R}^n$, $V = L^2(\Omega)$. We denote by $H_0^1(\Omega)$ the Sobolev space, that is the completion of the space $\mathcal{D}(\Omega)$ of infinitely differentiable functions with compact support for the norm $(\int_\Omega |\text{grad } x(\omega)|^2 \, d\omega)^{1/2}$. It is contained in $L^2(\Omega)$ and dense.

We set

(7) $$\phi(v) = \begin{cases} (1/2) \int_\Omega |\text{grad } x(\omega)|^2 \, d\omega & \text{if } x \in H_0^1(\Omega) \\ +\infty & \text{if } x \notin H_0^1(\Omega) \end{cases}.$$

Then we can compute that

(8) $$\phi^*(p) = (1/2) \int_\Omega |\text{grad } \Delta^{-1} p(\omega)|^2 \, d\omega$$

where the derivatives are distributions. Also

(9) $$\partial\phi(x) = \begin{cases} -\Delta x & \text{if } x \in H_0^1(\Omega) \cap H_0^2(\Omega) \\ \emptyset & \text{if } x \notin H_0^1(\Omega) \cap H_0^2(\Omega) \end{cases}.$$

Many linear and non-linear differential operators can be obtained in this way.

2. Elliptic equations

We consider a "potential" defined on a Hilbert space H by

(1) a lower semicontinuous convex function $\phi: V \to]-\infty,+\infty]$.

The purpose of the following lines is to characterize the solutions $x(\cdot)$ of the Dirichlet problem for the differential equation

(2) $$0 \in -\ddot{x} + \partial\phi(x); \quad x(0) = x(T) = 0.$$

It is well known that this is the Euler-Lagrange equations associated to the minimization of the Lagrangian

(3) $$L(x,\dot{x}) = \int_0^T [\phi(x,t)) + (1/2)\|\dot{x}(t)\|^2]dt$$

on the subspace $H_0^1 \equiv H_0^1(0,T;H)$ of functions $x(\cdot) \in H^1 \equiv H^1(0,T;H)$

vanishing at 0 and T. We recall that the Sobolev space H^1 is the subspace of functions $x \in L^2 \equiv L^2(0,T;H)$ whose weak derivatives $\dot{x} \in L^2$.

Let us make precise the variational principle characterizing the solutions of the elliptic differential equation (2).

PROPOSITION 2. A function $x \in H^2 \equiv H^2(0,T;H)$ is a solution to (2) if and only if

a) the pair $(x,\dot{x}) \in H_0^1 \times H^1$ minimizes on $H_0^1 \times H^1$ the energy functional defined by

$$A_E(y,q) = \int_0^T [\phi(y) + (1/2)||\dot{y}||^2 + \phi^*(\dot{q}) + (1/2)||q||^2]dt \tag{4}$$

and

b) $A_E(x,\dot{x}) = 0.$

We note that statement a) is equivalent to

a') $x \in H^1$ minimizes $\int_0^T [\phi(y) + (1/2)||\dot{y}||^2]dt$ on H_0^1

a") $\dot{x} \in H^1$ minimizes $\int_0^T [\phi^*(p) + (1/2)||p||^2]dt$ on H^1.

Proof. It depends upon the integration by parts formula. Indeed, since $q \in H^1$ and $y \in H_0^1$, we obtain

$$A_E(y,q) \geq \int_0^T [\langle\dot{q},y\rangle + \langle q,\dot{y}\rangle]dt = 0$$

So $A_E(x,p) = 0$ if and only if $\dot{p} \in \partial\phi(x)$ and $p = \dot{x}$, i.e. if and only if $\ddot{x} \in \partial\phi(x)$.

More generally we obtain analogous results for Lagrangians $L: L^2 \times L^2 \to]-\infty,+\infty]$ satisfying

(5) $\quad L$ is convex and lower semicontinuous.

We associate with this Lagrangian its Hamiltonian defined by

$$H(x,p) = \sup_v [\langle p,v\rangle - L(x,v)] \tag{6}$$

that satisfies

$$(7)\quad \begin{array}{ll} \text{i)}\ x \to H(x,p) & \text{is concave} \\ \text{ii)}\ p \to H(x,p) & \text{is convex and lower semicontinuous.} \end{array}$$

We also use the conjugate L^* of L and the energy functional A_E defined on $H_0^1 \times H^1$ by

$$(8)\qquad A_E(y,q) = L(y,\dot{y}) + L^*(\dot{q},q).$$

We also introduce the function K defined by

$$(9)\qquad K(r,v) = \sup_x [\langle r,x\rangle - L(x,v)]$$

that satisfies

$$(10)\quad \begin{array}{ll} \text{i)}\ r \to K(r,v) & \text{is convex and continuous} \\ \text{ii)}\ v \to K(r,v) & \text{is concave.} \end{array}$$

We define the function a_E by

$$(11)\qquad a_E(x,p;y,q) = K(\dot{p},\dot{y}) - K(q,x) + \int_0^T (\langle\dot{q},x\rangle) + \langle p,\dot{y}\rangle.$$

It is convex with respect to (x,p), concave with respect to (y,q) and satisfies $a_E(y,q;y,q) = 0$.

The minimization of $A_E(y,q)$ *on* $H_0^1 \times H^1$ *is, obviously, equivalent to the minimization of* $L(y,\dot{y})$ *on* H_0^1 *and the minimization of* $L^*(\dot{q},q)$ *on* H^1.

The latter problem is often called the *dual problem* of the former.

We say that a pair $(x,p) \in H_0^1 \times H_0^1$ is a solution to the *Euler-Lagrange equation* if

$$(12)\qquad (\dot{p},p) \in \partial L(x,\dot{x}).$$

Since $\partial L(x,\dot{x}) \subset \partial_x L(x,\dot{x}) \times \partial_v L(x,\dot{x})$, we find, by eliminating p, that x is a solution to

$$(13)\qquad 0 \in (d/dt)\partial_v L(x,\dot{x}) - \partial_x L(x,\dot{x}), \qquad u(0) = x(T) = 0.$$

We say that $(x,p) \in H_0^1 \times H^1$ solves the *Hamiltonian system* if

$$(14) \qquad \begin{cases} \text{i)} & \dot{p} \in \partial_x(-H)(x,p) \\ \text{ii)} & \dot{x} \in \partial_p H(x,p); \qquad x(0) = x(T) = 0 \end{cases}$$

THEOREM 1. The following conditions are equivalent:

a) The pair $(x,p) \in H_0^1 \times H^1$ is a solution to the Euler-Lagrange equations (12);

b) The pair $(x,p) \in H_0^1 \times H^1$ is a solution to the Hamiltonian system (14);

c^1) The pair $(x,p) \in H_0^1 \times H^1$ minimizes $A_E(y,q)$ on $H_0^1 \times H^1$

and c^2) $A_E(x,p) = 0$;

d) $(x,p) \in H_0^1 \times H^1$ is a solution to the variational inequalities

$$(15) \qquad a_E(x,p;y,q) \leq 0 \ \forall\, (y,q) \in H_0^1 \times H^1.$$

Proof. Equivalence between statements a) and b) follows from Proposition 1. We note that $A_E(x,p) = 0$ if and only if $(\dot{p},p) \in \partial L(x,\dot{x})$. Finally, we remark that

$$A_E(x,p) = \sup_{y,q} a_E(x,p;y,q).$$

So, to solve $A_E(x,p) = 0$ amounts to solving variational inequalities (15).

REMARK. Naturally, the condition c) is *redundant*.

In the first place, condition c^2) implies condition c^1). Also some suitable minimax theorems in infinite dimensional spaces imply some redundancy in condition c).

Assume for instance that

(16) a) $\operatorname{Dom} L = X \times Y$ where $X \subset H^1$ and $Y \subset L^2$

that

b) $0 \in \operatorname{Int}((d/dt)X - Y)$

and

c) that F is continuous at a point $(\tilde{x},\tilde{v}) \in L^2 \times L^2$ at least and that $x \to H(x,p)$ is upper semicontinuous.

Then if $x \in H_0^1$ minimizes the function $L(y,\dot{y})$ on H_0^1, there exists $p \in H^1$ such that either one of the equivalent conditions of Theorem 1 are satisfied. (See for instance J. P. Aubin [1], Theorem 14.3.2 of Chapter 14, Section 3.5 and I. Ekeland and R. Temam [9].)

This is true in particular if the minimum $x \in H_0^1$ of $L(y,\dot{y})$ is such that $(x,\dot{x}) \in \text{Int Dom } L$. Note, nevertheless, that the above assumptions imply the existence of $p \in H^1$ satisfying

(13) $$L^*(\dot{p},p) = \min_{q \in H} L^*(\dot{q},q) = -\inf_{y \in H'} L(y,\dot{y})$$

even if no minimum of $L(y,\dot{y})$ exists.

3. Parabolic Equations

We still consider the potential defined on a Hilbert space H by

(1) a lower semi-continuous convex function $\phi: V \to]-\infty,+\infty]$

and the solutions to the initial value problem for the differential equation

(2) $$0 \in \dot{x} + \partial\phi(x); \quad x(0) = 0.$$

The existence and uniqueness of a solution is well known (see Brézis [3]).

Brézis and Ekeland [4] characterized the solution of this differential equation by the following variational principle.

THEOREM 2. (Brézis-Ekeland [4]) A function $x \in H^1$ solves (2) if and only if

a) $x \in H^1$, $x(0) = 0$ minimizes on the space of functions $y \in H$ satisfying $y(0) = 0$ the function A_p defined by

(3) $$A_p(y) = \int_0^T [\phi(y) + \phi^*(-\dot{y})]dt + (1/2)\|y(T)\|^2$$

and b) $A_p(x) = 0$.

Proof: Indeed, if $y(0) = 0$, we obtain

$$A_p(y) \geq -\int_0^T \langle \dot{y},y\rangle dt + (1/2)\|y(T)\|^2 \geq 0$$

and

$$A_p(x) = 0 \quad \text{if and only if} \quad -\dot{x} \in \partial\phi(x).$$

We can go further and provide another characterization of the solution x of the parabolic equation (2). We define

$$(4) \qquad a_p(x,y) = \int_0^T [\phi^*(-\dot{x}) - \phi^*(-\dot{y})]dt + (1/2)\|x(T)\|^2 - \int_0^T \langle \dot{y},x\rangle dt.$$

This function is convex and lower semi-continuous with respect to x, concave and upper semi-continuous with respect to y and satisfies $a_p(y,y) = 0$. We prove:

PROPOSITION 3. A function $x \in H^1$ solves (2) if and only if

$$(5) \qquad \forall\, y \in H^1, \quad y(0) = 0, \quad a_p(x,y) \leq 0.$$

This result shows that minimax theorems can be used to prove the existence of a solution.

4. Hyperbolic Equations

Let always $\phi: H \to]-\infty,+\infty]$ be a lower semicontinuous convex function.

We consider the Dirichlet problem for the "hyperbolic" differential equation

$$(1) \qquad 0 \in \ddot{x} + \partial\phi(x); \quad x(0) = x(T) = 0.$$

The classical Newton equations as well as the wave equation belong to this class. If we regard this problem as a Euler-Lagrange equation, we see that the Lagrangian

$$L(x,v) = -\phi(x) + (1/2)\|v\|^2$$

is concave with respect to x and convex with respect to v. So, a solution x of (1) can no longer characterize the minimization of $L(x,\dot{x})$ on H^1_0. But we can construct another energy functional that can be used for devising a variational principle. We set

(2) $$A_H(y,q) = \int_0^T [\phi(y) + (1/2)\|q\|^2 + \phi^*(-\dot{q}) + (1/2)\|\dot{y}\|^2 - 2<q,\dot{y}>]dt.$$

PROPOSITION 4. A function $x \in H^2$ solves (1) if and only if

a) the pair $(x,\dot{x}) \in H^1_0 \times H^1$ minimizes on $H^1_0 \times H^1$ the energy functional $A_H(y,q)$

and

b) $A_H(x,\dot{x}) = 0$.

Proof. We check that

$$A_H(y,q) \geq \int_0^T [<-\dot{q},y> + <q,\dot{y}> - 2<q,\dot{y}>]dt = 0$$

so that $A(x,p) = 0$ if and only if $-\dot{p} \in \partial\phi(x)$ and $p = \dot{x}$, i.e. if and only if $-x \in \partial\phi(x)$.

We note that the Hamiltonian defined by

$$H(x,p) = \phi(x) + (1/2)\|p\|^2$$

is convex and lower semicontinuous with respect to both variables (x,p). This suggests the following generalization.

Let us assume that the Hamiltonian H satisfies

(3) $$H: L^2 \times L^2 \to]-\infty,+\infty]$$

is a lower semicontinuous convex function and we define the Lagrangian L by

(4) $$L(x,v) = \sup_p [<p,v> - H(x,p)].$$

It is concave with respect to x and convex with respect to v. We also introduce the conjugate function H^* of H and we set the energy function

$$(5) \qquad A_H(y,q) = H(y,q) + H^*(-\dot{q},\dot{y}) + \int_0^T (\langle\dot{q},y\rangle - \langle q,\dot{y}\rangle)dt.$$

THEOREM 3. A pair $(x,p) \in H_0^1 \times H^1$ is a solution to the Euler-Lagrange equation

$$(6) \qquad \begin{cases} \text{i)} & -\dot{p} \in \partial_x(-L)(x,\dot{x}) \\ \text{ii)} & p \in \partial_v L(x,\dot{x}); \qquad x(0) = x(T) = 0 \end{cases}$$

if and only if

a) the pair $(x,p) \in H_0^1 \times H^1$ minimizes $A_H(y,q)$ on $H_0^1 \times H^1$

and

b) $A_H(x,p) = 0$.

This is also equivalent to the fact that $(x,p) \in H_0^1 \times H^1$ solves the Hamiltonian system:

$$(7) \qquad (-\dot{p},\dot{x}) \in \partial H(x,p).$$

Proof. Again, it is quite obvious, since

$$A_H(y,q) \geq \int_0^T [\langle-\dot{q},y\rangle + \langle q,\dot{y}\rangle - 2\langle q,\dot{y}\rangle]dt = 0$$

and since

$$A_H(x,p) = 0 \text{ if and only if } (-\dot{p},\dot{x}) \in \partial H(x,p).$$

Since $v \mapsto L(x,v)$ is the conjugate function of $p \mapsto H(x,p)$, Proposition 1 implies that $p \in \partial_v L(x,\dot{x})$ and $-\dot{p} \in \partial_x(-L)(x,\dot{x})$. By eliminating p, we obtain the Euler-Lagrange equations.

5. Solving the Newton Equation

So, solving the Hamiltonian system (7) amounts to solving the equation

$$A_H(x,p) = 0.$$

For that purpose, we can make use again of the duality theory and write

$$A_H(x,p) = \sup_{y,q} a_H(x,p;y,q)$$

where

$$(8) \qquad a_H(x,p;y,q) = H^*(-\dot{p},\dot{x}) - H^*(-\dot{q},\dot{y}) + \int_0^T (\langle x,\dot{p}-\dot{q}\rangle - \langle p,\dot{x}-\dot{y}\rangle)dt\,.$$

PROPOSITION 5. The pair $(x,p) \in H_0^1 \times H^1$ is a solution to the Hamiltonian system (7) if and only if it solves the variational inequalities

$$(9) \qquad \forall\,(y,q) \in H_0^1 \times H^1, \quad a_H(x,p;y,q) \leq 0.$$

We note that the function a_H is lower semicontinuous with respect to (x,p) on bounded subsets and concave with respect to (y,q) and satisfies $a_H(y,q;y,q) \leq 0$.

Hence, we may use the Ky Fan inequality or its generalizations (Brézis-Nirenberg-Stampacchia) for solving variational inequalities (9). We recall Ky Fan's theorem (see J. P. Aubin [1], Chapters 7.1 and 13.1, for instance, of Ky Fan [10]).

THEOREM 4. Let us assume that X is a convex compact subset (of a Hausdorff locally convex vector space) and that $a: X \times X \to \mathbb{R}$ satisfies

$$(10) \qquad \begin{cases} \text{i)} & \forall\, y \in X, \quad x \mapsto a(x,y) \text{ is lower semicontinuous} \\ \text{ii)} & \forall\, x \in X, \quad y \mapsto a(x,y) \text{ is concave} \\ \text{iii)} & \forall\, y \in X, \quad a(y,y) \leq 0. \end{cases}$$

Then there exists $x \in X$ such that

$$(11) \qquad \forall\, y \in X, \quad a(x,y) \leq 0.$$

We can apply Ky Fan's theorem when X is a ball of radius n of $H_0^1 \times H^1$ and when a_H is defined by (8). For instance, we obtain the following theorem (see Aubin-Ekeland [2]).

THEOREM 5. Let the potential function $\phi: H \to \mathbb{R}$ be convex continuous and satisfy the following growth condition;

(12) $\quad \exists\, k > 0, \quad \exists\, c > 0$ such that $\phi(x) \leq k\|x\|^2 + c \;\forall\, x \in H.$

Then there exists $T_k > 0$ such that problem (1) has at least one solution whenever $T \in]0,T_k[$. Moreover $T_k \to \infty$ when $k \to 0$.

6. The Principle of Least Action and Its Dual

We see that the functional $A(y,q)$ can be written as the sum of the functionals

(1) $$M(y,q) = H(y,q) + \int_0^T \langle \dot{q},y\rangle dt$$

and

(2) $$E(y,q) = H^*(-\dot{q},\dot{y}) - \int_0^T \langle q,\dot{y}\rangle dt.$$

The principle of least action (Maupertuis' principle) states that any pair $(x,p) \in H_0^1 \times H^1$ minimizing $M(y,q)$ is a solution to the Hamiltonian system.

In the framework of periodic solutions to the Hamiltonian system, Clarke and Ekeland show a kind of dual principle:

PROPOSITION 6. Let us assume that $(x,p) \in H_0^1 \times H^1$ minimizes $E(y,q)$ on $H_0^1 \times H^1$ and that $(-\dot{p},\dot{x}) \in \operatorname{Int} \operatorname{Dom} H^*$. Then (x,p) is a solution to the Hamiltonian system.

Proof. If (x,p) minimizes $E(y,q)$, we have for all $\theta > 0$,

$$0 \leq (H(-\dot{p}+\theta(-\dot{q}),\dot{x}+\theta\dot{y}) - H(-\dot{p},\dot{x}))/\theta - \int_0^T (\langle q,\dot{x}\rangle + \langle p,\dot{y}\rangle)dt$$
$$- \theta \int_0^T \langle q,\dot{y}\rangle dt.$$

Since $(-\dot{p},\dot{x}) \in \operatorname{Int} \operatorname{Dom} H^*$, H^* has a directional derivative from the right; so

$$0 \leq \inf_{(y,q)\,\varepsilon\, H_0^1 \times H} \left[DH^*(-\dot{p},\dot{x})(-\dot{q},\dot{y}) - \int_0^T (<q,\dot{x}> + <p,\dot{y}>)dt\right]$$

We recall that $DH^*(-\dot{p},\dot{x})\ (\cdot,\cdot)$ is the support function of the non-empty weakly compact convex subset $\partial H^*(-\dot{p},\dot{x})$. So, we obtain

$$0 \leq \inf_{(y,q)\,\varepsilon\, H_0^1 \times H^1} \ \sup_{(r,v)\,\varepsilon\, \partial H^*(-\dot{p},\dot{x})} \int_0^T [<\dot{q},x - r> + <v - p,\dot{y}>]dt.$$

By the lopsided minimax theorem (see J. P. Aubin [1], Chapter 7, §1), we deduce that there exists $(\bar{r},\bar{v})\ \varepsilon\ \partial H^*(-\dot{p},\dot{x})$ such that

$$0 \leq \inf_{(y,q)\,\varepsilon\, H_0^1 \times H^1} \int_0^T [<\dot{q},x - \bar{r}> + <\bar{v} - p,\dot{y}>]dt.$$

Hence $\bar{r} = x$, $\bar{v} = p$ and thus, $(x,p)\ \varepsilon\ \partial H^*(-\dot{p},\dot{x})$. This is equivalent to $(-\dot{p},\dot{x})\ \varepsilon\ \partial H(x,p)$. This ends the proof.

By using this method, Clarke and Ekeland proved that for all T, there exists a periodic solution (x,p) of the Hamiltonian system with minimal period T, when $H(x,p)$ behaves like $(|x|^2 + |p|^2)^\alpha$, $1/2 < \alpha < 1$. When $H(x,p)$ behaves like $(|x|^2 + |p|^2)^\alpha$, $\alpha > 1$, P. Rabinowitz proved the existence of a non-trivial solution of period T. Ekeland proved an analogous result by using the variational approach. Either one of these results implies the existence of a periodic solution whose orbit lies in $H^{-1}(h)$ for all h.

Naturally, other functionals have the property that their minimum (x,p) are solutions to the Hamiltonian system. For instance, Ekeland and Lasry [8] introduced the functional

$$EL(y,q) = H(-\dot{q},\dot{y} - (H(-q,y) + \alpha) - \beta<q,\dot{y}>.$$

They used it to prove that, under suitable assumptions (non-resonance), there exist at least n distinct periodic solutions of the Hamiltonian system on $H^{-1}(h)$.

CONCLUSIONS

It is easy to note that other variational principles fully characterizing solutions of differential equations can be devised.

Also, we can replace the differential operator d/dt by any kind of differential operator Λ: We replace the integration by part by the Green formula relation Λ and its formal adjoint.

In the same way, many kinds of boundary conditions can be treated this way.

The fact that variational problems of the form (2-15), (3-5), and (5-8) are equivalent to differential equations is useful for proving existence of solutions, in the sense that the function a involves the *derivatives* of the functions (x,p).

REFERENCES

1. J. P. Aubin, *Mathematical Methods of Game and Economic Theory*, North Holland, 1979.

2. J. P. Aubin and I. Ekeland, Second-order evolution equations associated with convex Hamiltonians, *Can. Math. Bull.*, 1979.

3. H. Brézis, *Opérateurs maximaux monotones et semi-groupes de contractions dans les espaces de Hilbert*, North Holland.

4. H. Brézis and I. Ekeland, Un principe variationnel associé à certaines équations paraboliques, *C.R.A.S. 282:* (1976) 971-974 and 1197-1198.

5. F. H. Clarke, Periodic solutions to Hamiltonian inclusions, to appear.

6. F. H. Clarke and I. Ekeland, Hamiltonian trajectories having prescribed minimal period, *Comm. Pure Appl. Math.*: (1979).

7. I. Ekeland, Periodic solutions of Hamiltonian equations and a theory of P. Rabinowitz, *J. Diff. Equations*: (1979).

8. I. Ekeland and J. M. Lasry, Nombre de solutions périodiques des équations d'Hamilton, *Cah. Math.*: December (1979).

9. I. Ekeland and R. Temam, *Convex analysis and variational problems*, North Holland, 1976.

10. Ky Fan, A minimax inequality and applications. In Shishia (Ed.), *Inequalities III*, Academic Press, 1972, 103-113.

11. M. Schatzmann, A class on non-linear differential equations of second order in time, *Non-linear Analysis 2*: (1978) 355-373.

CONVEXITY AND DIFFERENTIABILITY IN BANACH SPACES: STATE OF THE ART

N. Ghoussoub
University of British Columbia
C.E.R.E.M.A.D.E.
Université de Paris IX Dauphine

INTRODUCTION

For the last ten years, after the innovative work of Asplund in 1967, the differentiability of functions defined on infinite dimensional spaces, were subject to an extensive and deep study by a large number of mathematicians. In these notes, we were interested in surveying--without giving any proof--the various (topological, geometrical and measure theoretical) characterizations of Asplund spaces; that is the spaces in which every continuous convex function is Frechet differentiable on a dense G_δ of its domain.

In the first section, we deal with the extremal structure of the dual of an Asplund space. A complete and excellent survey on this part are the lecture notes of Phelps at University College, London [26].

The second section is concerned with the Radon-Nikodym property which is already a well-understood subject. The recent book of Diestel and Uhl [5] provides a readable and wideranging treatment of the subject, while the duality between R.N.P. and Asplund spaces can also be found in Phelps' lecture notes.

The third section summarizes the remarkable results of Stegall in [29], [30], while most of the renorming results can be found in the book of Diestel [4] and the recent paper of Davis, Ghoussoub and Lindenstrauss [6].

For the Banach spaces with local unconditional structure, we refer to the notes of Lacey [22] and the new characterizations of Asplund spaces with l.u.s.t. can be found in [13] and [14].

The list of references that we give here is far from being complete. For a more complete bibliography, we refer to the books and lecture notes mentioned above.

I. Differentiability of Convex Functions and Asplund Spaces

Let E denote a real Banach space, D a non-empty open convex subset of E and ϕ a convex function on D.

DEFINITION I.1.

a) A functional x^* in E^* is said to be a *subdifferential* of ϕ at $x_0 \in D$ if it satisfies

$$<x^*, x - x_0> \leq \phi(x) - \phi(x_0) \quad \forall\, x \in D.$$

b) The set of all subdifferentials of ϕ at x_0 is denoted by $\partial\phi(x_0)$ and is called the subdifferential of ϕ.

An immediate consequence of Hahn-Banach theorem and Alaoglu's theorem is

PROPOSITION I.2. If ϕ is convex and continuous, then for each $x \in D$, the set $\partial\phi(x)$ is non-empty, convex and weak-star compact. Moreover, the subdifferential map $x \to \partial\phi(x)$ is locally bounded, that is: for any $x_0 \in D$, there exists $M > 0$ and a neighborhood U of x_0 such that $||x^*|| \leq M$ whenever $x^* \in \partial\phi(x)$ and $x \in U$.

DEFINITION I.2. A continuous convex function ϕ is *Gateaux differentiable* at $x_0 \in D$ if the limit

$$d\phi(x_0)(x) = \lim_{t \to 0} (\phi(x_0 + tx) - \phi(x_0))/t$$

exists for each $x \in E$. The function $d\phi(x_0)$ is then a linear continuous functional on E and is called the *Gateaux* differential of ϕ at x_0.

PROPOSITION I.4. A continuous convex function ϕ is Gateaux differentiable at x if and only if $\partial\phi(x)$ is a single point.

EXAMPLES I.5.

a) The norm $\|x\|_1 = \sum_n |x_n|$ in $\ell_1(\mathbb{N})$ is Gateaux differentiable exactly at those points $x = (x_n)$ in ℓ_1 for which $x_n \neq 0$ for all n. The differential is then the bounded sequence $(sg_n x_n)_n \in \ell_\infty$.

b) The norm in $\ell_1(\Gamma)$ (Γ uncountable) is not Gateaux differentiable at any point.

DEFINITION I.6. A continuous function ϕ is said to be *Fréchet* differentiable at $x_0 \in D$, if the limit

$$\phi'(x_0)(x) = \lim_{t \to 0} (\phi(x_0 + tx) - \phi(x_0))/t$$

is uniform as $t \to 0$ for $\|x\| \leq 1$.

EXAMPLES AND REMARKS I.7.

a) In finite dimensional spaces, the local lipschitz property and the compactness of the unit ball implies that Gateaux and Fréchet differentiability coincide.

b) The norm in $\ell_1(\mathbb{N})$ is not Fréchet differentiable at any point.

c) The square of the norm in a Hilbert space is everywhere Fréchet differentiable and $\phi'(x)$ is given by the functional $y \to 2\langle x,y\rangle$.

The most striking results on differentiability in finite dimensional spaces are those which assert that "reasonable" functions are differentiable on "fat" sets. For instance, a well known classical result asserts that a convex function on an open interval D of $\mathbb{R}$, is differentiable but on a countable subset of D. A deeper result is the theorem of Rademacher which shows that locally lipschitz functions on $\mathbb{R}^n$ are differentiable almost everywhere. In the latter we are only concerned with the convex case.

DEFINITION I.8. A Banach space E will be called an *Asplund* space [*Weak Asplund*] if every continuous convex function on an open convex subset of E is Fréchet differentiable [Gateaux differentiable] in a dense G_δ subset of its domain.

We have already seen that ℓ_1 is not an Asplund space and $\ell_1(\Gamma)$ is not Weak Asplund if Γ is uncountable. More generally Mazur in 1933 had already proven

PROPOSITION I.9. Every separable Banach space is a Weak Asplund space.

The following theorem gives the connection between Fréchet differentiability and the continuity of the differential map.

THEOREM I.10.

a) If ϕ is Fréchet differentiable at x_0, then the subdifferential map $x \to \partial\phi(x)$ is norm to norm upper semi-continuous at x_0: That is if $x_n \in D$, $\|x_n - x_0\| \to 0$ and $x_n^* \in \partial\phi(x_n)$, then $\|x_n^* - \phi'(x_0)\| \to 0$.

b) Conversely, if ϕ is Gateaux differentiable at x_0 and $x \to \partial\phi(x)$ is norm to norm upper semi-continuous at x_0, then ϕ is Fréchet differentiable at x_0.

It is worth noting that the subdifferential map is always norm to weak-star upper semi-continuous. In particular, if the Gateeaux differential $d\phi(x_0)$ exists, then $x_n^* \to d\phi(x_0)$ weak-star whenever $x_n^* \in \partial\phi(x_n)$ and $\|x_n - x\| \to 0$.

The natural class of continuous convex functions associated to a Banach space E is the class of Gauge (or Minkowski) functions associated with convex subsets of E with non-empty interior. Recall that if C is such a set, then

$$P_C(x) = \inf\{\lambda > 0;\ x \in \lambda C\} \quad \text{for any } x \text{ in } E$$

is positive-homogeneous, subadditive, non-negative and continuous. Moreover

$$C^0 = \{x^* \in E^*;\ \langle x^*, x\rangle \leq 1 \quad \text{for all } x \in E\}$$

which is the polar of C, is weak-star compact and convex and $x^* \in C^0$ if and only if $\langle x^*, x\rangle \leq P_C(x)$ for all $x \in E$.

It turns out that the study of differentiability properties of continuous convex functions can often be reduced to the study of such properties for continuous gauge functionals. The crucial part is that for the latter functions, differentiability can be reformulated in terms of extremal properties of the corresponding polar sets.

DEFINITION I.11. If f is a continuous linear functional on a Banach space and C is a bounded subset of the space, denote by $M(f,C)$ the supremum of f on C.

a) We say that f *exposes* x in C if X is the unique point of C such that $f(x) = M(f,C)$. A point of C is an *exposed point* if there is some f which exposes it.

b) The functional f *strongly exposes* $x \in C$ if f exposes x and $||x_n - x|| \to 0$ whenever $(x_n) \subseteq C$ and $f(x_n) \to f(x)$. A point is called *strongly exposed point* if there exists such a functional f.

c) If $\alpha > 0$, the set $S(f,C,\alpha) = \{x \in C;\ f(x) > M(f,C) - \alpha\}$ is called a *slice* of C. A point x of C is a *denting point* if for every $\varepsilon > 0$, there exists a slice of C of diameter less than ε and containing x. It is readily seen that

$$\text{Extreme}(C) \supseteq \text{Exposed}(C) \supseteq \text{Strongly Exposed}(C)$$

and

$$\text{Extreme}(C) \supseteq \text{Denting}(C) \supseteq \text{Strongly Exposed}(C) \ .$$

The following proposition is the first connection between differentiability and extremal properties.

PROPOSITION I.12. Let p be a continuous gauge functional on E, with $C = \{x;\ p(x) \leq 1\}$. Then p is Gauteaux [Fréchet] differentiable at x with differential x^*, if and only if $x^* \in C^0$ and x weak-star exposes [strongly exposes] C^0 at x^*.

The following theorem which refines the above proposition summarizes the work of many authors following the lead of E. Asplund [1].

For details, we refer to Phelps [26].

THEOREM I.13. For a Banach space E, the following properties are equivalent:

1) E is an Asplund space.
2) Every weak-star compact convex non-empty subset of E^* is the weak-star closed convex hull of its weak-star strongly exposed points.
3) Each such a set contains at least one weak-star strongly exposed point.
4) Every non-empty bounded subset of E^* admits weak-star slices of arbitrarily small diameter.
5) Every equivalent norm on E has at least one point of Fréchet differentiability.
6) Every continuous gauge functional on E has at least one point of Fréchet differentiability.

It is worth comparing 2) to the classical Krein-Milman theorem applied to the weak-star compact convex of E^*. The above theorem shows also that it is enough to study the differentiability of continuous gauge functionals and conclude for all convex continuous functions defined on the space. We shall see in Section IV that it is actually enough to have one equivalent norm which is Fréchet differentiable everywhere (but at zero).

II. Differentiability of Vector Measures and Radon-Nikodym Spaces

Let E be a Banach space and let $(\Omega,\mathcal{F},P)$ be a probability space.

DEFINITION II.1. A function $f\colon \Omega \to E$ is said to be *Pettis-integrable* if

(i) for every $x^* \in E^*$, the map $\omega \to x^*f(\omega)$ is P-measurable and P-integrable.

(ii) for every $A \in \mathcal{F}$, there is x_A in E such that $x^*(x_A) = \int_A x^*f(\omega)dP(\omega)$ for every x^* in E^*. In this case we write $x_A = \text{Pettis} - \int_A f dP$.

DEFINITION II.2. A function $f: \Omega \to E$ is said to be *Bochner integrable* if there exists a sequence (f_n) of step function such that

(i) $\lim_n \|f(\omega) - f_n(\omega)\| = 0$ for P-almost all ω in Ω.

(ii) $\lim_n \int_\Omega \|f(\omega) - f_n(\omega)\| dP(\omega) = 0$.

We define Bochner $- \int_A fdP = \lim_n \int_A f_n dP$.

DEFINITION II.3. A Banach space has the *Radon-Nikodym property* (R.N.P.) [resp. the *weak Radon-Nikodym property* (W.R.N.P.)] if for every probability space $(\Omega,\mathcal{F},P)$ and every vector measure $m: \mathcal{F} \to E$ such that $\|m(A)\| \leq P(A)$ $(\forall\ A \in \mathcal{F})$, there exists a Bochner integrable (resp. Pettis-integrable) function $f: \Omega \to E$ such that $m(A) = \text{Bochner} - \int_A fdP$ [resp. $m(A) = \text{Pettis} - \int_A fdP$] $\forall\ A \in \Sigma$.

The Radon-Nikodym property has been studied extensively in the last ten years. For a complete study, we refer to the book of Diestel and Uhl [5]. In the following theorem we summarize the most important geometrical characterizations of this property.

THEOREM II.4. For a Banach space E, the following properties are equivalent:

1) E has the R.N.P.
2) Every bounded subset of E is subset dentable (has slices of arbitrarily small diameter).
3) Every closed bounded convex subset of E is the norm closed convex hull of its denting points.
4) Every closed bounded convex subset of E is the norm closed convex hull of its strongly exposed points.

The characterization 4) is due to Phelps [27], while 2) and 3) are due to Rieffel [28], Maynard [23], Huff [17] and Davis-Phelps [7]. A main tool was the martingale characterization of the R.N.P. established by Chatterji [3].

It is clear that the R.N.P. implies the Krein-Milman Property, that is: every closed bounded convex subset is the closed convex hull of its extreme points. The converse is still a longstanding open question.

The following theorem shows the remarkable duality between differentiability of functions and that of measures. It is too bad that the connection is not direct, but via the extremal properties inherited by the subsets of the dual space.

THEOREM II.5. For a Banach space E, the following conditions are equivalent:

1) E is an Asplund space.
2) E^* has the Radon-Nikodym property.
3) E^* has the Krein-Milman property.
4) Every separable subspace of E has a separable dual.

That 3) $\Leftrightarrow$ 4) was shown by Huff and Morris [18] using the construction employed by Stegall [29] when he showed that 4) $\Leftrightarrow$ 2). That 4) $\Leftrightarrow$ 1) was proved by Stegall [29]. Another proof is due to Namioka who uses the following characterization of Asplund spaces, worth comparing to 4) of Theorem I.13.

PROPOSITION II.6. A Banach space E is Asplund if and only if every non-empty weak-star compact subset of E^* contains non-empty weak-star relatively open subsets of arbitrarily small diameter.

The dual characterization of Asplund spaces is a very handy tool for identifying these spaces, since sometimes it is much easier to verify the R.N.P. in Banach spaces. Immediate applications are the following stability properties.

COROLLARY II.7.

a) Every closed subspace of an Asplund space is Asplund.
b) A finite product of Asplund spaces is Asplund.
c) If F is a closed subspace of an Asplund space E, then E/F is Asplund. Equivalently, any continuous linear *onto* image of an Asplund space is an Asplund space.
d) If F and E/F are Asplund spaces, then E is an Asplund space.

An enlightening example is that it is not that straightforward that $E \times \mathbb{R}$ is an Asplund space whenever E is, while a measure valued in $F \times G$ has trivially a Radon-Nikodym density whenever it has partial ones in F and G.

COROLLARY II.8.

a) Every reflexive space is an Asplund space.

b) $c_0(\Gamma)$ is an Asplund space.

c) If E^* is weakly compactly generated [Weakly $\mathcal{K}$-analytic, or weakly Lindelöf] then E is an Asplund space.

d) $C(K)$ is an Asplund space if and only if K is scattered. Namioka-Phelps [25].

For the proofs of c) we refer to Talagrand [31] and Edgar [10].

III. Asplund Operators

Most of the results of this section are due to Stegall [29], [30].

DEFINITION III.1.

(i) A subset S of $L_\infty(\Omega,\Sigma,\mu)$ is said to be *equimeasurable* [15], if for every $\varepsilon > 0$, there exists a set B in Σ $\mu(B) > \mu(\Omega) - \varepsilon$ and $\{f1_B;\ f \in S\}$ is a relatively compact subset of $L_\infty(\Omega,\Sigma,\mu)$.

(ii) A bounded subset S of a Banach space E is said to be a *G.S.P. set* if for any finite measure space (Ω,Σ,μ) and any continuous linear function $T: E \to L_\infty(\Omega,\Sigma,\mu)$, the set $\{Tx;\ x \in S\}$ is equimeasurable.

The introduction of the above notion was motivated by a definition of Grothendieck [15]. It turns out that the class of G.S.P. sets has very nice permanence properties, and is dual in some sense to the class of dentable sets (Radon-Nikodym sets).

PROPOSITION III.2.

a) S is a G.S.P. set if and only if the smallest closed absolutely convex subset of E containing S is G.S.P.

b) The collection of G.S.P. subsets of E forms a hereditary ring and the sum of two G.S.P. sets is G.S.P.

c) S is G.S.P. if and only if every separable subset is G.S.P.

d) If (S_n) is a sequence of G.S.P. sets, (λ_n) a sequence of reals going to zero

$$\bigcap_{n=1}^{\infty} [S_n + B(0,\lambda_n)] \text{ is G.S.P.}$$

e) If $T: E \to F$ is a bounded linear operator and S is a G.S.P. set, so is $T(S)$.

f) Let (E_n) be a sequence of Banach spaces and S a bounded subset of $E = (\Sigma \oplus E_n)_{c_0}$. Then S is a G.S.P. set if and only if $P_n(S)$ is a G.S.P. set for every n; here P_n denotes the canonical operator from E to E_n.

DEFINITION III.3. A continuous linear operator $T: E \to F$ is said to be an *Asplund operator* if for any $\phi: F \to \mathbb{R}$ that is continuous and convex, ϕT is differentiable on a dense subset of E.

By adapting the factorization technique of Davis-Figiel-Johnson and Pelcynski [8], Stegall proved the following:

THEOREM III.4. Let $T: E \to F$ be a continuous linear operator. Then the following are equivalent:

1) T is an Asplund operator.

2) The image of the unit ball of E by T is a G.S.P. set.

3) The image of the unit ball of F^* by T^* is a dentable set (T^* is a radon-Nikodym operator).

4) T factors through an Asplund space. That is there exists an Asplund space G, two operators $T_1: E \to G$ and $T_2: G \to F$ such that the following diagram commutes:

$$\begin{array}{ccc} E & \xrightarrow{\;T\;} & F \\ & {\scriptstyle T_1}\searrow \quad \nearrow{\scriptstyle T_2} & \\ & G & \end{array}$$

5) T^* factors through a space with the Radon-Nikodym property.

Now we have a few topological characterizations of Asplund spaces.

THEOREM III.5. A banach space E is Asplund if and only if every bounded subset of E is G.S.P.

We finish this section with the remarkable dichotomy theorem for the factorization of operators between Banach spaces.

THEOREM III.6. If $T: E \to F$ is linear bounded operator then one and only one of the following is true:

(i) T is an Asplund operator.

(ii) T is a factor of the Haar operator $H: \ell_1 \to L_\infty(\Delta,\mu)$.

That is there exists $T_1: \ell_1 \to E$ and $T_2: F \to L_\infty(\Delta,\mu)$ such that the following diagram commutes. For the definition of the Haar operator we refer to [29].

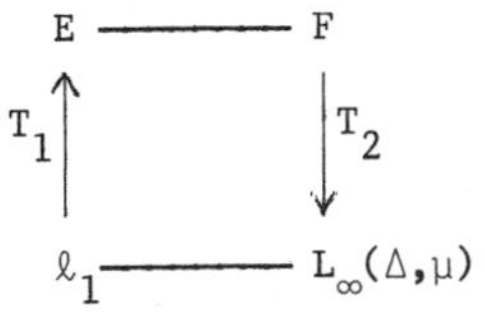

IV. Renorming Asplund Spaces

In Theorem I.13, we mentioned a characterization of Asplund spaces in terms of the existence of a Fréchet differentiability point for every equivalent norm on the space. The question then arises, if the differentiability (everywhere!) of one equivalent norm is enough to force the space to be Asplund. And what about the converse?

The first question was answered affirmatively by Ekeland and Lebourg [9].

THEOREM IV.1. If there exists on a Banach space E an equivalent norm which is Fréchet differentiable everywhere (but at zero) then E is an Asplund space.

Another property for norms closely connected to Fréchet differentiable ones is the following:

DEFINITION IV.2. A norm $\|\ \|$ on a Banach space E is said to be *locally uniformly convex* if whenever (x_n) is a sequence such that $\|x_n + x\| \to 2\|x\|$ and $\|x_n\| \to \|x\|$ then $\|x_n - x\| \to 0$.

It is immediate to see that for such norms, the weak and the norm topologies coincide on the unit sphere of E. The following straightforward proposition gives the connection with Asplund spaces.

PROPOSITION IV.3. If the dual norm on E^* is locally uniformly convex then the norm on E is Fréchet differentiable.

The theorem of Ekeland-Lebourg implies then the following:

COROLLARY IV.4. If there exists an equivalent *dual* norm on E^* which is locally uniformly convex then E is an Asplund space (and E^* has R.N.P.).

Recently, the converse of the above corollary was shown untrue by the construction of the long James space [11] which is a dual of an Asplund space, but cannot be renormed dually in a locally uniformly convex manner.

The converse to Theorem IV.2 is still unsettled, since it is still unkonwn if the existence of an equivalent Fréchet differentiable norm on E implies the existence of an equivalent locally uniformly convex norm in the dual E^*. However, John and Zizler [21] proved the following:

THEOREM IV.5. If E is W.C.G. (weakly compactly generated) and if there exists on E an equivalent norm which is Fréchet differentiable, then there exists on E an equivalent norm which is Fréchet differentiable, locally uniformly convex and such that its dual norm is locally uniformly convex.

Their proof consists of a refinement of Trojanski's renorming of W.C.G. spaces [32] and the application of Asplund's Averaging technique [1]. John and Zizler [20] proved the same conclusion of Theorem IV.5 provided E and E^* are W.C.G. This was an extension of a theorem of Kadec-Klee for spaces with separable duals [4].

In the case of Banach lattices, the following converse to the theorem of Ekeland-Lebourg was proved in [6], [12]:

THEOREM IV.6. An order complete Banach lattice E is Asplund if and only if there exists an equivalent Fréchet differentiable, locally uniformly convex lattice norm on E such that its dual norm is locally uniformly convex.

V. Asplund Spaces with Local Unconditional Structure

The most interesting characterizations of a property in an abstract Banach space are in terms of the embedding of classical Banach spaces into that space. By the remarks of Section I, it is clear that if E is an Asplund space, then E does not contain a subspace isomorphic to ℓ_1 (we say ℓ_1 does not embed in E). Unfortunately, the converse need not hold and the counterexample is the famous James tree space [19]. However, for a large class of spaces, which includes Banach lattices, spaces with unconditional basis and the $\mathcal{L}_p$ spaces, the converse can be shown to be true.

One characterization (that we may use for a definition) of a space with local unconditional structure (l.u.s.t.) is the property that their duals are complemented in a Banach lattice.

THEOREM V.1. For a Banach space E with l.u.s.t. the following are equivalent:

1) E is an Asplund space.

2) ℓ_1 does not embed in E.

3) L_1 does not embed in E^*.

4) E^* has the weak Radon-Nikodym property.

5) c_0 does not embed in E^* and L_1 does not embed complementably in E^*.

6) Every weak-star compact convex subset of E^* is the norm closed convex hull of its extreme points.

2) $\Leftrightarrow$ 3) $\Leftrightarrow$ 4) $\Leftrightarrow$ 6) hold in any Banach space. See for instance Musial [24] and in the case of l.u.s.t. y space 4) $\Leftrightarrow$ 1) was proved by Ghoussoub-Saab [13] and 3) $\Leftrightarrow$ 5) by Figiel-Ghoussoub-Johnson [14].

A consequence of the above and a theorem of Hagler-Stegal [16] is the following:

COROLLARY V.2. If E has l.u.s.t. $(\ell \oplus \ell_\infty^n)_{\ell_1}$ does not embed in E and c_0 does not embed in E^* then E is an Asplund space.

Another characterization which can be found in Figiel-Ghoussoub-Johnson [14] is the following:

THEOREM V.3. If E is complemented in a Banach lattice, $C(\Delta)$ does not embed in E and c_0 does not embed in E^* then E is an Asplund space.

FINAL REMARKS

While the structure of Asplund space is, by now, well understood, not much is known on Gateaux differentiability spaces. Other than the weak Asplund spaces, one may introduce the *Gateaux differentiability spaces* (every convex continuous function defined on a convex open set D is Gateaux differentiable on a dense subset of D) and the spaces of *Minkowski differentiability* (every continuous gauge function is Gateaux differentiable on a dense subset). The structure of these spaces is still a complete mystery. Even the most elementary stability properties are still open questions. For a list of related problems we refer to Phelps [26].

REFERENCES

1. E. Asplund, Fréchet differentiability of convex functions, *Acta Math. 121:* (1968) 31-47.

2. E. Asplund and T. Rockafellar, Gradients of convex functions, *Trans. Amer. Math. Soc. 139:* (1969) 433-467.

3. S. D. Chatterji, Martingale convergence and the Radon-Nikodym theorem in Banach spaces, *Math. Scand. 22:* (1968) 21-41.

4. J. Diestel, Topics in the geometry of Banach spaces, *Springer-Verlag Lecture Notes 485:* (1975).

5. J. Diestel and J. Jr. Uhl, Vector measures, *Amer. Math. Soc. Surveys 15:* (1977).

6. W. Davis, N. Ghoussoub, and J. Lindenstrauss, A lattice renorming theorem and applications to vector valued processes, *Trans. Amer. Math. Soc. 263* (2): (1981) 531-540.

7. W. Davis and R. Phelps, The Radon-Nikodym property and dentable sets in Banach spaces, *Proc. A.M.S. 45:* (1974).

8. W. Davis, T. Figiel, W. Johnson and A. Pekynski, Factoring weakly compact operators, *Journal of Funct. Analysis 17:* (1974) 311-327.

9. I. Ekeland and G. Lebourg, Generic Fréchet differentiability and perturbed optimization problems in Banach space, *Trans. Amer. Math. Soc. 224:* (1976) 193-216.

10. G. Edgar, Measurability in a Banach space, *Indiana Journal of Math. 26:* (1977) 663-677.

11. G. Edgar, A long James space, to appear (1979).

12. N. Ghoussoub, Renorming dual Banach lattices, to appear *Proc. of the A.M.S.:* (1980).

13. N. Ghoussoub and E. Saab, On the weak Radon-Nikody property, *Proc. A.M.S. 81* (1): (1981) 81-84.

14. T. Figiel, N. Ghoussoub, and W. Johnson, On the structure of non weakly compact operators on Banach lattices, to appear (1981).

15. A. Grothendick, Produits tensoriels et espaces nucleaires, *Memoires A.M.S. 16:* (1955).

16. J. Hagler and C. Stegall, Banach spaces whose duals contain complemented subspaces isomorphic to C]0,1], *J. Functional Analysis 13:* (1973) 233-251.

17. R. Huff, Dentability and the Radon-Nikodym property, *Duke Math. Journal 41:* (1974) 111-114.

18. R. Huff and P. Morris, Dual spaces with the Krein-Milman property have the Radon-Nikodym property, *P.A.M.S. 49:* (1975) 104-108.

19. R. James, A separable somewhat reflexive space with non-separable dual, *Bulletin A.M.S. 80:* (1974) 738-743.

20. K. John and V. Zizler, A renorming of dual spaces, *Israel J. Math. 12:* (1972) 331-336.

21. K. John and V. Zizler, A note of renorming of dual spaces, *Bull. Acad. Polon. Sci. 21:* (1973) 47-50.

22. E. Lacey, Local unconditional structure in Banach spaces, *Springer-Verlag Lecture Notes 604:* (1977) 44-56.

23. H. Maynard, A geometric characterization of Banach spaces possessing the Radon-Nikodym theorem, *Trans. A.M.S. 185:* (1973) 493-500.

24. K. Musian, The weak Radon-Nikodym property, *Studia Math. 64:* (1978) 151-174.

25. J. Namioka and R. Phelps, Banach spaces which are Asplund spaces, *Duke Math. J. 42:* (1975) 735-749.

26. R. Phelps, Differentiability of convex functions on Banach spaces, *Lectures Notes at University College London:* (1978).

27. R. Phelps, Dentability and extreme points in Banach spaces, *J. of Funct. Analysis 16:* (1974) 78-90.

28. M. Rieffel, The Radon-Nikodym theorem for the Bochner integral, *Trans. Amer. Math. Soc. 131:* (1968) 466-487.

29. C. Stegall, The Radon-Nikosym property in conjugate Banach spaces, *Trans. Amer. Math. Soc. 206:* (1975) 213-223.

30. C. Stegall, The Radon-Nikodyn property in conjugate spaces II, preprint.

31. M. Talagrand, Sur la structure borélienne des espaces analytiques, *Bull Sci. Math. 101:* (1977) 415-422.

32. S. Trojanski, On locally uniformly convex and differentiable norms in certain non-separable Banach spaces, *Studia Math. 37:* (1971) 173-180.

PARTIAL DIFFERENTIAL EQUATIONS AND CONTROL

SOME UNILATERAL PROBLEMS OF THE NON-VARIATIONAL TYPE

G. M. TROIANIELLO

Universita di Roma

C.E.R.E.M.A.D.E.
Université de Paris IX Dauphine

ABSTRACT

We study existence, uniqueness and regularity of solutions of elliptic unilateral problems for operators not of divergence form. We also study the convergence of the ergodic means of solutions of parabolic problems to the above.

RÉSUMÉ

On étudie l'existence, l'unicité et la régularité des solutions de problèmes unilatéraux elliptiques pour des opérateurs non sous forme divergence. On étudie aussi la convergence des moyens ergodiques des solutions de problèmes paraboliques analogues aux précédents.

INTRODUCTION

This paper surveys unilateral problems for *elliptic* operators of *non-variational* [or *non-divergence*] form. It grows out of a previous article [20], but almost all the results stated here are elliptic versions of their *parabolic* counterparts proved in [21]. We have sketched the proofs of some of these results whenever, in addition to a straightforward passage from evolution operators to stationary ones, some changes (although minor) were needed.

The necessity of such changes was twofold. On the one hand, here we also consider the case of *discontinuous* coefficients (of the *Cordes type*), whereas in [21] only the *continuous* case was considered. On the other hand, regularization of non-variational operators leads to variational inequalities with *non-coercive* bilinear forms, (a situation of some importance in elliptic problems, not in parabolic ones). It can also be added that the ergodic theorem of the last section appears to be new.

Most of the material presented here was illustrated at the "Colloquium du C.E.R.E.M.A.D.E." directed by J. M. Lasry, under the auspices of the cooperation agreement between the Universities of Paris IX Dauphine and Rome.

1. Formulation of the Problem

1.1 The problem we are interested in is the following: To find a function u defined on a given bounded open subset Ω of R^N such that

$$(UP)\qquad \begin{cases} u = 0 \quad \text{on} \quad \partial\Omega, \\ \\ u < \psi, \quad Lu \leq f, \quad (Lu - f)(u - \psi) = 0 \quad \text{in} \quad \Omega. \end{cases}$$

In (UP) $\partial\Omega$ is the boundary of Ω, which we assume is of class C^2, and the data L, ψ and f are fixed, at the outset, as follows: L is a second-order linear differential operator,

$$L \equiv - \sum_{i,j=1}^{N} a_{ij}(\partial^2/\partial x_i \partial x_j)$$

(we do without lower-order terms only for the sake of simplicity), with

$$(1.1)\qquad a_{ij} \in L^\infty(\Omega) \qquad \forall\ i,j = 1,\dots,N$$

and

$$(1.2)\qquad \sum_{i,j=1}^{N} a_{ij}\xi_i\xi_j \geq \alpha|\xi|^2 \qquad \text{a.e. in } \Omega,\ \forall\ \xi \in R^n\ (\alpha > 0);$$

(1.3) $\psi \in L^2(\Omega)$;

(1.4) $f \in L^2(\Omega)$.

with assumption (1.1) replaced by the stronger one

(1.5) $a_{ij} \in C^o(\bar{\Omega}) \quad \forall\ i,j = 1,\dots,N,$

problems such as (UP) appear in the theory of stochastic control, and have been investigated with probabilistic tools by Bensoussan and Lions in §7 of [2]; cf. also [21].

The *bilateral* problem analogous to (UP) will be the object of a forthcoming paper by M. G. Garroni and M. A. Vivaldi.

1.2 As a preliminary step to our investigation of (UP) let us determine an appropriate formulation of the corresponding *unconstrained* problem: To find a function u defined on Ω such that

(P) $\bar{u} = 0$ on $\partial\Omega$, $\quad L\bar{u} = f$ in Ω

(that is formally, the same problem as (UP) except for (1.3) replaced by $\psi \equiv +\infty$).

Since L maps the Sobolev space $H^2(\Omega)$ continuously into $L^2(\Omega)$, the natural function-theoretical formulation of (P) is

(1.6) $\bar{u} \in H_o^1(\Omega) \cap H^2(\Omega), \quad L\bar{u} = f \quad$ a.e. in Ω,

For $N \geq 3$, however, under the sole assumptions (1.1) and (1.2) the Dirichlet problem (1.6) is not *well-posed* (in the sense of Hadamard), as the following example shows (cf. [19], [13]). Let Ω be the ball of radius R around the origin, and let

$$a_{ij}(x) \equiv \delta_{ij} + \Big((N+\lambda-2)/(1-\lambda)\Big)\Big(x_i x_j / x^2\Big)$$

for $x \neq 0, \quad \forall\ i,j = 1,\dots,N,$

with $1 > \lambda > 2 - N/2$.

Then (1.1) and (1.2) hold; at the same time, both functions

$$\bar{u} \equiv 0$$

and

$$\bar{u} \equiv |x|^{\lambda} - R^{\lambda}$$

satisfy (1.6) with

$$f \equiv 0.$$

We are therefore led to strengthen our assumptions about the coefficients of L. This we do by requiring (1.1) to be either replaced by (1.5) or implemented by

$$(1.7) \qquad \Big(\sum_{i=1}^{N} a_{ii}\Big)^2 \Big(\sum_{i,j=1}^{N} a_{ij}^2\Big)^{-1} \geq N - \varepsilon \quad \text{a.e. in } \Omega \quad (\varepsilon < 1)$$

(condition of the Cordes types), (1.2) still valid.

Then the following can be ascertained. First of all (cf. [19], [7], [8]), problem (1.6) admits a unique solution $\bar{u}$, which satisfies

$$(1.8) \qquad \|\bar{u}\|_{H^2(\Omega)} \leq C\|f\|_{L^2(\Omega)},$$

with C dependent on the modulus of uniform continuity of the a_{ij}'s in case (1.5), on ε in case (1.1), (1.7), as well as on Ω and N. In other terms, L is a continuous isomorphism $H_o^{\prime}(\Omega) \cap H^2(\Omega) \xrightarrow{\text{onto}} L^2(\Omega)$. Actually (cf. [6],[7]), L is an analogous isomorphism $H_o^{\prime}(\Omega) \cap H^{2,p}(\Omega) \xrightarrow{\text{onto}} L^p(\Omega)$ for all exponents p in certain intervals, and precisely: in the case (1.5) $\forall\, p \in]1,+\infty[$, in the case (1.1), (1.7) $\forall\, p \in]p_0,p_1[$ with p_0 and p_1 suitable real numbers (generally *unknown*) satisfying

$$1 < p_0 < 2 < p_1 < +\infty.$$

Thus in the case (1.5), as a consequence of Sobolev's imbedding theorems, $\bar{u}$ is Hölder continuous on $\bar{\Omega}$ provided

$$(1.9) \qquad p > N/2 .$$

In the case (1.1), (1.7) no specific use can be made of Sobolev's theorems for lack of information about the magnitude of p_1; with a different approach, however, Hölder continuity of $\bar{u}$ can again be proven (cf. [9],[8]), under the assumption

(1.10) $\quad p > 2N/3.$

Finally, in both cases (1.5) and (1.1), (1.7) L satisfies the maximum principle (cf. [7],[8]).

1.3 The preceeding considerations suggest the following formulation of problem (UP).

Assume either (1.5) of (1.1), (1.7), as well as (1.2), (1.3), (1.4); a solution to the problem

$$(1.11)\quad \begin{cases} u \in H_0^1(\Omega) \cap H^2(\Omega), \\ u \leq \psi, \quad Lu \leq f, \quad (Lu - f)(u - \psi) = 0 \quad \text{a.e. in } \Omega \end{cases}$$

is then sought for.

2. Some Remarks About Non-Coercive Variational Inequalities

2.1 Under assumption (1.1), even if replaced by (1.5) or implemend by (1.7), L cannot be put into variational form. The situation changes if we replace (1.1) by:

(2.1) $\quad a_{ij} \in C^1(\bar{\Omega}) \qquad \forall\ i,j = 1,\dots,N,$

(1.2) still valid, *as we assume throughout the present section.* Then, L maps $H^1(\Omega)$ continuously into the dual $H^{-1}(\Omega)$ of $H_0^1(\Omega)$; a continuous bilinear form $a(v,w)$ on $H_0^1(\Omega)$ is defined by setting

$$a(v,w) \equiv \langle Lv,w\rangle \qquad \text{for } v,w \in H_0^1(\Omega)$$

(with $\langle\cdot,\cdot\rangle \equiv$ the duality pairing between $H^{-1}(\Omega)$ and $H_0^1(\Omega)$, so that

$$a(v,w) = \sum_{i,j=1}^{N} \int_\Omega \left[a_{ij}\left(\frac{\partial v}{\partial x_i}\right)\left(\frac{\partial w}{\partial x_j}\right) + \left(\frac{\partial a_{ij}}{\partial x_j}\right)\left(\frac{v}{x_i}\right) w \right] dx .$$

A simple computation shows that there exist two constants $\mu > 0$

and $\gamma \geq 0$ such that

$$(2.2) \qquad a(v,v) \geq \mu\|v\|^2_{H'(\Omega)} - \gamma\|v\|^2_{L^2(\Omega)} \qquad \forall\ v \in H_0'(\Omega).$$

Thus, $a(v,w)$ is not *coercive* on $H_0'(\Omega)$ unless $\gamma = 0$, which is not true in general.

If a solution u to (1.11) (with the present choice of L) exists, then v solves the *variational inequality*

$$(2.3) \qquad u \in K(\psi), \qquad a(u,v-u) \geq \int_\Omega f(v-u)dx \ \forall\ v \in K(\psi),$$

where

$$(2.4) \qquad K(\psi) \equiv \{v \in H_0'(\Omega) \mid v \leq \psi \ \text{ a.e. in } \Omega\}.$$

Viceversa, if (2.3) has a solution u *which moreover belongs to* $H^2(\Omega)$, then (1.11) also admits u as a solution.

These circumstances (easily checked by use of the Green formula) demonstrate the relevance to our problem of (2.3), insofar as *existence, uniqueness* and *regularity* of solutions are concerned.

Since $a(v,w)$ is not coercive of $H_0'(\Omega)$, the basic existence and uniqueness result of Stampacchia [18], Lions-Stampacchia [14] cannot be directly invoked. We can, instead, proceed as follows. We set

$$(2.5) \qquad \hat{\Phi}_L(\psi,f) \equiv \{w \in H'(\Omega) \mid w \leq 0 \ \text{ on } \partial\Omega,\ w \leq \psi \ \text{ a.e. in } \Omega$$
$$Lw \leq f \ \text{ in the sense of } H^{-1}(\Omega)\};$$

then we have

LEMMA 2.1. Assume (1.3), (1.4) together with

$$(2.6) \qquad K(\psi) \neq \emptyset$$

and

$$(2.7) \qquad \hat{\Phi}_L(\psi,f) \neq \emptyset.$$

Then a solution of (2.3) exists.

LEMMA 2.2. Assume (1.3), (1.4). A solution of (2.3), and a fortiori a solution of (1.11), if existing, satisfies

$$(2.8) \qquad u = \sup \Phi_L(\psi,f),$$

and is therefore unique.

Proof of Lemma 2.1. We follow an iteration schema as in the existence part of Theorem 3.1.5 of [2]. Let $u^o \equiv \bar{u}$ (the solution of (1.6)), and let u^n, $n \in N$, be defined by

$$(2.9) \qquad u^n \in K(\psi), \quad a_\gamma(u^n, v-u^n) \geq \int_\Omega (f+u^{n-1})(v-u^n)dx \quad \forall\, v \in K(\psi),$$

where

$$a_\gamma(v,w) \equiv \langle (L+\gamma)v,w\rangle \quad \text{for} \quad v,w \in H_0^1(\Omega).$$

The form $a_\gamma(v,w)$ being coercive on $H_0^1(\Omega)$ as a consequence of (2.2), there exists a unique to (2.9). By induction, the maximum principle for L and the comparison theorem for solutions of coercive variational inequalities yield

$$(2.10) \qquad u^o \geq u^n \geq u^{n+1} \geq w \qquad \text{a.e. in } \Omega, \ \forall\, n \in N,$$

where w is any fixed element of $\Phi_L(\psi,f)$.

From (2.10) it follows that the sequence $\{u^n\}_{n\in N}$ is bounded in $L^2(\Omega)$, and hence, by (2.9) (and (2.2)), in $H_0^1(\Omega)$. Let u be the weak limit of $\{u^n\}_{n\in N}$ in $H_0^1(\Omega)$ (no need to pass to a subsequence, thanks to the monotonicity ensured by (2.10)). Because of the weak lower semicontinuity of the positive bilinear form $a_\gamma(v,w)$, (2.9) yields

$$u \in K(\psi), \quad a_\gamma(u,v-u) \geq \int_\Omega (f+\gamma u)(v-u)dx \quad \forall\, v \in K(\Omega),$$

which is nothing (2.3).

Lemma 2.2, of fundamental importance to our purposes, has been proven in [15]; for a different proof of uniqueness, with (1.3) and

(1.4) replaced by

$$\psi \in L^{\infty}(\Omega)$$

and

$$f \in L^{\infty}(\Omega)$$

respectively, see the uniqueness part of the above-mentioned theorem of [2].

Let us now pass to a regularity result. Most of such results (e.g. see [4]) can be transferred from the coercive to the non-coercive case through a simple "bootstrap" argument. In the sequel we shall specifically need the following easy extension of the Lewy-Stampacchia inequality (cf. [20]).

LEMMA 2.3. Let u solve (2.3) with

$$\psi \equiv \bigwedge_{k=1}^{m} \psi^{(k)}, \qquad \psi^{(k)} \in H^2(\Omega),$$

(1.4) still valid. Then u satisfies

$$Lu \geq \left(\bigtriangleup_{k=1}^{m} L\psi^{(k)} \right) \wedge f \quad \text{in the sense of } H^{-1}(\Omega)$$

in addition to

$$Lu \leq f \quad \text{in the sense of } H^{-1}(\Omega);$$

hence, in particular,

$$u \in H^2(\Omega), \tag{2.11}$$

so that u solves (1.11) too.

2.2 Each of the three lemmas above holds true under more general assumptions about the coefficients of L (and about ψ, f).

On the other hand, it will be of use in the sequel to take into account the limits of their scope, as illustrated by the following counterexamples (which actually concern a coercive case).

First of all, it may occur that the solution of (2.3) does not satisfy (2.11), even for a rather regular (e.g., Lipschitz continuous) obstacle. This is shown by

COUNTEREXAMPLE 2.1. Let $N = 1$, $\Omega \equiv]-1,1[$, $L \equiv -d^2/dx^2$, $\psi(x) \equiv |x| - 1/2$, $f \equiv 0$. Then, the function

$$u(x) \equiv (1/2)(|x| - 1)$$

solves (2.3); its second derivative in the sense of distributions is not even a measurable function on (it is the Dirac measure concentrated at the origin).

Secondly, it could happen (2.7) is satisfied, whereas (2.6) is not--and hence (2.3) is meaningless. This can be seen in

COUNTEREXAMPLE 2.2. Let N, Ω, L and f be as in Counterexample 2.1, and let $\psi(x) \equiv -1|x|)^{1/2}$. The set $\Phi_L(\psi,f)$ then consists of convex function in $H'(\Omega)$ which lie below ψ, so that (2.7) holds true; (2.8) *defines* a function u, which indeed coincides with $\psi \notin H_0'(\Omega)$.

2.3 Part of the above can be summarized as follows:

$$\begin{array}{l} u \quad \text{solves} \quad (1.11) \\ \quad\ \Downarrow \ \not\Uparrow \\ u \quad \text{solves} \quad (2.3) \\ \quad\ \Downarrow \ \not\Uparrow \\ u \quad \text{satisfies} \ (2.8), \end{array}$$

so that we may consider the second and the third stage as weak formulations of (UP), the latter weaker than the former.

3. Existence and Uniqueness of Solutions

3.1 Let us return to our problem (1.11) as formulated in subsection 1.3. Since L is now of the non-variational type, we can no longer prove the existence of a solution to (1.11) *via* the existence of a solution to a variational inequality plus a regularity result. We can however approximate L by a sequence of operators for which such a procedure is available. This we do next.

THEOREM 3.1. Together with (1.2), assume either (1.5) or (1.1), (1.7). Also assume

$$\psi \equiv \bigwedge_{k=1}^{m} \psi^{(k)}, \quad \psi^{(k)} \in H^2(\Omega), \quad \psi^{(k)} \geq 0 \quad \text{on} \quad \partial\Omega.$$

Finally, assume (1.4). Then problem (1.11) has a solution u, which satisfies also

$$Lu > \left(\bigwedge_{k=1}^{m} L\psi^{(k)}\right) \wedge f \qquad \text{a.e. in} \quad \Omega.$$

Proof. We only consider the slightly more difficult case (1.1), (1.7). After normalizing L so to get

$$\text{(3.1)} \qquad \sum_{i=1}^{N} a_{ii} = 1 \qquad \text{a.e. in} \quad \Omega,$$

let each function a_{ij} be extended to all of R^N by setting

$$a_{ij} = \delta_{ij} \qquad \text{in} \quad R^n \backslash \Omega,$$

and let

$$a^\eta_{ij}(x) \equiv \eta^{-N} \int_{R^N} ((x-y)/\eta) a_{ij}(y) dy \quad \forall x \in R^N,$$

where $\eta > 0$ and $\zeta \in C_0^\infty(R^N)$, with $\zeta \geq 0$ in R^N, $\zeta(x) = 0$ for $|x| \geq 1$, and $\int_{R^N} \zeta dx = 1$. Thus, $\forall i,j = 1,\ldots,N$ $a^\eta_{ij} \in C_0^\infty(R^N)$, $a^\eta_{ij} \to a_{ij}$ in $L^q(\Omega)$ $\forall q \in [1,+\infty[$; moreover, the a^η_{ij}'s share the same properties (1.1), (1.2), (1.7) and (3.1) as the a_{ij}'s (see [19] for what concerns (1.7)). Let

$$L^\nu \equiv - \sum_{i,j=1}^{N} a_{ij}(\partial^2/\partial x_i \partial x_j) .$$

We are in a position to apply both Lemma 2.1 (since the solution

w to the problem

$$w \in H_0^1(\Omega) \cap H^2(\Omega), \qquad L^\eta w^\eta = \left(\bigwedge_{k=1}^{m} L^\eta \psi^{(k)}\right) \wedge f \qquad \text{a.e. in } \Omega,$$

belongs to $K(\psi)$ as well as to $\Phi_{L^\eta}(\psi,f)$ thanks to the maximum principle) and Lemma 2.3, with L replaced by $L^{\eta L}$. Thus, the analog for L^η of (2.3) has a unique solution $u^\eta = \sup \Phi_{L^\eta}(\psi,f)$ which belongs to $H^2(\Omega)$ and satisfies

$$(3.2) \qquad f \geq L^\eta u^\eta \geq \left(\bigwedge_{k=1}^{m} L^\eta \phi^{(k)}\right) \wedge f \qquad \text{a.e. in } \Omega.$$

Because of (3.2), $\|L^\eta u^\eta\|_{L^2(\Omega)}$ is bounded by a constant independent of $\eta > 0$, and hence (1.8) implies that the same is true of $\|u^\eta\|_{H^2}$. Let u be the limit of a subsequence of $\{u^\eta\}_{\eta>0}$ weakly convergent in $H^2(\Omega)$ (Ω); an easy computation shows that u has all the required properties.

3.2 Passing to uniqueness, let us now prove the analog of Lemma 2.2. Since we are in the non-variational case, $\Phi_L(\psi,f)$ given by (2.5) is not well-defined; we replace it by

$$(3.3) \qquad \phi_L(\psi,f) \equiv \{w \in H^2(\Omega) \mid w \leq 0 \text{ on } \partial\Omega,\ w \leq \psi \text{ a.e. in } \Omega,\ Lw \leq f \text{ a.e. in } \Omega\}.$$

THEOREM 3.2. Under the same assumptions of Theorem 3.1 the coefficients of L and f, together with (1.3), a solution of (1.11), if existing, satisfies

$$(3.4) \qquad u = \sup \phi_L(\psi,f)$$

and is therefore unique.

Proof. Let us again restrict ourselves to the case (1.1), (1.7); in the case (1.5) the proof is exactly the same as for the corresponding parabolic operator [21].

With L^η as in the preceeding proof, we have

$$\begin{cases} u \in H_0^1(\Omega) \cap H^2(\Omega), \\ u \le \psi, \quad L^\eta u \le f + (L^\eta - L), \quad u, \quad (L^\eta u - (f + (L^\eta - L)u)(u - \psi) = 0 \\ \qquad \text{a.e. in } \Omega, \end{cases}$$

whence, by Lemma 2.2 applied to L^η, $f + (L^\eta - L)u$,

(3.5) $$u = \sup \Phi_L(\psi, f + (L^\eta - L)u) = \sup \phi_{L^\eta}(\psi, f + (L^\eta - L)u).$$

Now let $v \in \phi_L(\psi,f)$, and let v^η solve the Dirichlet problem

$$v^\eta - v \in H_0^1(\Omega) \cap H^2(\Omega), \quad L^\eta v^\eta = Lv - |(L^\eta - L)v| - |(L^\eta - L)u| \quad \text{a.e. in } \Omega.$$

Then, it is easily checked that

$$v^\eta \in \Phi_{L^\eta}(\psi, f + (L^\eta - L)u),$$

so that (3.5) yields

(3.6) $$v^\eta \le u \quad \text{a.e. in } \Omega.$$

On the other hand, both $|(L^\eta - L)v|$ and $|(L^\eta - L)u|$ tend to 0 in any space $L^p(\Omega)$ with $p \in [1,2[$, hence in particular with $p \in]p_0, 2[$, p_0 being the number defined in subsection 1.2. For such p, v^η converges in $H^{2,p}(\Omega)$ toward the solution $\tilde{v}$ to the problem

$$\tilde{v} - v \in H_0^1(\Omega) \cap H^{2,p}(\Omega), \quad L\tilde{v} = Lv \quad \text{a.e. in } \Omega,$$

whence $\tilde{v} = v$. Then, (3.6) yields the inequality

$$v \le u \quad \text{a.e. in } \Omega,$$

from which (3.4) follows.

3.3 The above can be partly summed up, also in the light of subsection 2.2, in the following scheme:

$$u \quad \text{solves} \quad (1.11)$$
$$\Downarrow \; \not\Uparrow$$
$$u \quad \text{satisfies} \; (3.4),$$

which enables us to consider the second stage as a weak formulation of (UP). The intermediate stage of the analogous scheme in subsection (2.3) (the variational inequality) is, of course, not available for operators of the non-variational type, nor, as already pointed out, for operators of the variational type itself, if (2.6) is not satisfied.

This motivates the approach adopted in the next section.

4. Generalized Solutions of Problem (UP)

4.1 Assume either (1.5) or (1.1), (1.7), together with (1.2), about the coefficients of L; (1.3) about ψ; (1.4) about f. The set $\phi_L(\psi,f)$ defined by (3.3) admits a least upper bound in $L^2(\Omega)$ as soon as it satisfies

$$(4.1) \qquad \phi_L(\psi,f) \neq \emptyset,$$

(See [17], Theorem 4.9.) We introduce the notation

$$(4.2) \qquad \sigma_L(\psi,f) \; \sup \phi_L(\psi,f),$$

with the convention

$$\sigma_L(\psi,f) \equiv -\infty$$

if (4.1) does not hold.

LEMMA 4.1. (i) If $\sigma_L(\psi,f)$ exists in $L^2(\Omega)$, then $\sigma_L(\sigma_L(\psi,f),f)$ also does, and coincides with the former; (ii) The mapping $\psi,f \mapsto \sigma_L(\psi,f)$ is increasing and concave on $L^2(\Omega) \times L^2(\Omega)$; (iii) If $\psi = \psi_1$ for $i - 1,2$ are such that $\psi_1 - \psi_2 \in L^\infty(\Omega)$, then $\sigma_L(\psi_1,f)$ exists in $L^2(\Omega)$ if $\sigma_L(\psi_2,f)$ does, and the function $\sigma_L(\psi_1,f) - \sigma_L(\psi,f)$ belongs to $L^\infty(\Omega)$ with

$$\|\sigma_L(\psi_1(,f) - \sigma_1(\psi_2,f)\|_{L^\infty(\Omega)} \leq \|\psi_1 - \psi_2\|_{L^\infty(\Omega)}.$$

Properties (i) and (ii) follow immediately from the definition (4.2); as for (iii), its proof is exactly the same as in the evolution case [21].

4.2 Thanks to Theorem 3.2, Lemma 4.1 lists expected, and useful properties of the solution to (1.11) whenever it exists. Since, however, the Lemma basically amounts to *all* that can be said about the function defined by (4.2) under assumptions (1.3) and (1.4), the latter appear too general from the point of view on information content.

Let us therefore proceed to strengthen them. We replace (1.3) by

$$(4.3) \qquad \psi \in C^o(\bar{\Omega}), \qquad \psi \geq 0 \quad \text{on} \quad \partial\Omega,$$

and (1.4) by

$$(4.4) \qquad f \in L^p(\Omega),$$

with either (1.9) or (1.10) depending on the choice of (1.5) or (1.1), (1.7), respectively. Let us explicitly remark that, because of (4.3), assumption (4.1) is always fulfilled. We have the following:

THEOREM 4.1. Under the above assumptions, $\sigma_L(\psi,f)$ belongs to $C^o(\bar{\Omega})$, as well as to $H^2(\Omega')$ whenever Ω' is an open subset of Ω with a C^2 boundary such that

$$(4.5) \qquad \bar{\Omega}' \subseteq D^\psi \equiv \{x \in \bar{\Omega} | \ \sigma_1(\psi,f)(x) < \psi(x)\}.$$

Moreover, $\sigma_L(\psi,f)$ satisfies

$$(4.6) \qquad \sigma_L(\psi,f) = 0 \ \text{ on } \ \partial\Omega, \ \ \sigma_L(\psi,f) \leq \psi \ \text{ on } \ \bar{\Omega}, \ \ L\sigma_L(\psi,f) = f \ \text{ a.e. in } \ D^\psi.$$

Proof. As in [21], we approximate ψ in $C^o(\Omega)$ by a sequence $\{\psi^n\}_{n\in N}$ of smooth functions, in the present case

$$(4.7) \qquad \psi^n \in H^{2,p}(\Omega), \qquad \psi^n \geq 0 \quad \text{on } \partial\Omega,$$

the exponent p in (4.7) being the same as in (4.4). Then (by

Theorems 3.1 and 3.2),

$$u^n \equiv \sigma_L(\psi^n, f)$$

solves (1.11) with ψ replaced by ψ^n, and hence satisfies

$$f \geq Lu^n \geq (L\psi^n) \wedge f \quad \text{a.e. in } \Omega.$$

Because of the regularity results mentioned in subsection 1.2, each u^n is Hölder continuous on $\bar{\Omega}$; since (iii) of Lemma (4.1) yields

$$\|\sigma_L(\psi^n, f) - \sigma_1(\psi, f)\|_{L^\infty(\Omega)} \leq \|\psi^n - \psi\|_{L^\infty(\Omega)} \quad \forall n \in N,$$

$\sigma_L(\psi, f)$ is continuous on $\bar{\Omega}$ and vanishes on $\partial\Omega$.

The behavior of $\sigma_L(\psi, f)$ in the set D^ψ defined in (4.5) is a slightly more delicate matter. Indeed, it is clear that, if Ω' satisfies (4.5), then there exists an open set Ω'' of the same type such that $\text{dist}(\Omega', \Omega \setminus \Omega'') > 0$, and

$$\bar{\Omega}'' \subseteq D^{\psi^n} \quad \forall n \geq \bar{n},$$

for a suitable natural number $\bar{n}$. Thus,

(4.8) $\quad Lu^n = f \quad \text{a.e. in } \Omega'' \ \forall n \geq \bar{n}.$

The crucial passage in the conclusion of the theorem is that (4.8) leads to the estimate

$$\|u^n\|_{H^2(\Omega')} \leq \text{constant independent of } n \geq \bar{n}.$$

Such an estimate can be proven analogy with the corresponding classical Schauder estimate, *via* a passage from Hölder function spaces to Sobolev spaces. In the parabolic case, this procedure is outlined in [12], Ch. IV, §10.

4.3 Theorem 4.1 justifies called $\sigma_L(\psi, f)$ the *generalized solution* of problem (UP).

As we know from Theorem 3.2, the solution of (1.11), if existing, is the generalized solution of (UP). It would be desirable to prove that the generalized solution of (UP), if (existing and) sufficiently regular, solves (1.11). At the moment, this we are able to do only under assumptions (4.3), (4.4), by an almost straightforward application of Lemma 4.1 (i) (cf. [21]).

LEMMA 4.2. Under the same assumptions of Theorem 4.1, let $\sigma_L(\psi,f) \in H^2(\Omega)$. Then $u \equiv \sigma_L(\psi,f)$ solves (1.11).

5. Implicit Unilateral Problems

5.1 Let us now consider the situation arising when the abstacle ψ, instead of being kept fixed, "varies with the solution u". More precisely, we want to deal with an *implicit unilateral problem* such as

$$\text{(IUP)} \qquad \begin{cases} u = 0 \quad \text{on} \quad \partial\Omega, \\ u \leq Mu, \quad Lu \leq f, \quad (Lu - f)(u - M(u)) = 0 \quad \text{in} \quad \Omega, \end{cases}$$

instead of (UP); in (UP), M is a non-linear mapping between function spaces.

When dealing with operators of the variational type, the weak formulation of (IUP) given by (2.3) with $K(\psi)$ replaced by $K(M(u))$ is called a *quasi-variational inequality*; see [1], [3], [11].

Specifically, we suppose that

(j) M is a continuous, increasing and concave mapping $C^o(\bar{\Omega}) \to C^o(\bar{\Omega})$,

and that

(jj) there exists $\underline{u} \in C^o(\bar{\Omega})$ and $\delta \in]0,1[$ such that

$$\begin{cases} \underline{u} \leq \bar{u} \quad \text{on} \quad \bar{\Omega}, \quad M(\underline{u}) > 0 \quad \text{on} \quad \partial\Omega, \\ (1-\delta)\underline{u} + \delta\bar{u} < \sigma_L(M(\underline{u}),f) \quad \text{in} \quad \Omega, \end{cases}$$

with u given by (1.6), f satisfying (4.4) (same distinction on the magnitude of p as in section 4).

A *generalized solution* of (IUP) is a fixed point of the mapping

$$S \equiv \sigma_L(\cdot,f) \circ M,$$

which clearly shares the same properties (j) as M thanks to Lemma 4.1 and Theorem 4.2. Thus, such a generalized solution, if existing, solves (IUP) in the sense of Theorem 4.1, with ψ replaced by $M(u)$.

Exactly as in the evolution [21], existence and uniqueness follow from a suitable adaptation of the method employed in [10] for quasi-variational inequalities. We thus have

THEOREM 5.1. Under the same assumptions concerning L and f as in Theorem 4.1, there exists a unique function $u \in C^o(\bar{\Omega})$, $u \geq \underline{u}$, such that

$$u = S(u).$$

Moreover, given any $u^o \in C^o(\bar{\Omega})$ such that

$$u^o > \underline{u} \quad \text{on } \bar{\Omega},$$

u is the limit in $C^o(\bar{\Omega})$ of the sequence $\{S^n(u^o)\}_{n\epsilon N}$, with the following rate of convergence:

$$\|S^n(u^o) - u\|_{L^\infty(\Omega)} \leq (1-\delta)^n \|\bar{u} - \underline{u}\|_{L^\infty(\Omega)},$$

given by (j).

5.2 There are cases when the generalized solution to (IUP) actually solves its strong formulation, namely:

$$(5.1) \quad \begin{cases} u \in H_0^1(\Omega) \cap H^2(\Omega), \\ u \leq M(u), \quad Lu \leq f, \quad (Lu - f)(u - M(u)) = 0 \quad \text{a.e. in } \Omega. \end{cases}$$

An example pertaining to *systems of* implicit unilater problems can be found in [20]. Another example concerns the case when

$$(5.2) \qquad M(\phi)\ (x) \quad 1 + \bigwedge_{\substack{\xi \geq 0 \\ x+\xi \in \bar{\Omega}}} \phi(x+\xi) \ \forall\, x \in \bar{\Omega}.$$

In the variational case, the quasi-variational inequality for (5.2) was introduced by Bensoussan and Lions [1] to characterize the solution of a stochastic impulse control problem.

With an adaptation of the variational method due to Caffarelli-Friedman [5], Mosco [16], the following regularity result can be proved:

THEOREM 5.2. Let the following hold true:

$$\partial\Omega \ \text{ is of class } \ C^{2,\alpha},$$

$$a_{ij} \in C^{o,\alpha}(\bar{\Omega}) \ \forall\, i,j = 1,\dots,N,$$

(1.2) being still valid, and

$$f \in C^{o,\alpha}(\bar{\Omega}),$$

for a suitable $\alpha \in\ ]0,1[$. Moreover, let $\underline{u}$ be given by

$$\underline{u} \equiv \sigma_L(o,f)$$

satisfy

$$\underline{u} > -1 \ \text{ on } \ \bar{\Omega}.$$

Then the generalized solution to (IUP) relative to (5.2) belongs to $H^{2,p}(\Omega)$ $\forall\, p \in [1,+[$, and hence, in particular, satisfies (5.1).

As for its parabolic analog, the proof of this theorem makes crucial use of the fact that, since (4.6) holds true (with $\psi = M(u)$), u belongs to $C^{2,\alpha}(D^{M(u)})$ (cf. [13]).

6. An Ergodic Theorem

6.1 Let us recall a result pertaining to parabolic unilateral problems of the non-variational type [21] which we shall utilize in the proof of our next theorem.

Assume (1.5), with (1.2) still valid, together with (1.4), and let

(6.1) $\psi \in H^2(\Omega), \quad \psi \geq 0 \quad \text{on} \quad \partial\Omega,$

(6.2) $u_o \in H_0^1(\Omega) \cap H^2(\Omega), \quad u_o \leq \psi \quad \text{a.e. in} \quad \Omega.$

For each $T \in \,]0,+\infty[$ let

$$Q_T \equiv \Omega \times]0,T[, \quad \Sigma_T \equiv \partial\Omega \times]0,T[.$$

The parabolic unilateral problem

$$(6.3) \quad \begin{cases} u_T \in L^2(0,T,H_o^1(\Omega) \cap H^2(\Omega) \quad \text{with} \quad (\partial u/\partial t)T \in L^2(Q_T), \\ u_T(0) = u_o \quad \text{a.e. in} \quad \Omega, \\ u_T \leq \psi, \; (\partial u_T/\partial t) + Lu_T \leq f, \end{cases}$$

$$((\partial u_T/\partial t) + Lu_T - f)(u_T - \psi) = 0 \quad \text{a.e. in} \quad Q_T$$

has a unique solution u_T, which moreover satisfies

(6.4) $(\partial uT/\partial t) + Lu_T \geq (L\psi) \wedge f \quad \text{a.e. in} \quad Q_T.$

(the assumptions of the above being by no means the most general ones under which such a result holds true).

6.2 We are interested in the behavior as $T \to +\infty$, of the average

(6.5) $Av(u_T) \equiv (1/T) \int_0^T u_T(t)dt;$

the right-hand side of (6.5) is a Bochner integral which belongs to $H_o^1(\Omega) \cap H^2(\Omega)$.

THEOREM 6.1. Let ψ be fixed according to 6.1), and let u_o be any function satisfying (6.2). Then, as $T \to +\infty$, $Av(u_T)$ converges weakly in $H^2(\Omega)$, and hence strongly in $H_o^1(\Omega)$, toward the solution u of (1.11).

Proof. By integrating from 0 to T the second inequality on the third line of (6.3) as well as (6.4) we obtain

(6.6) $$f \geq \left((u_T(T) - u_o)/T\right) + L\, Av(u_T) \geq (L\psi) \wedge f \quad \text{a.e. in } \Omega.$$

Let us prove that

(6.7) $$\|u_T(T)\|_{L^2(\Omega)} \leq \text{Constant independent of } T \in]0,+\infty[.$$

Indeed, $u_T(T)$ satisfies

(6.8) $$u_T(T) \leq \psi \quad \text{a.e. in } \Omega, \ \forall\, T \in]0,+\infty[.$$

On the other hand, $u_T(T)$ also satisfies

(6.9) $$u_T(T) \geq \phi \quad \text{a.e. in } \Omega, \ \forall\, T \in]0,+\infty[,$$

where ϕ is any function such that

(6.10) $$\phi \in \zeta_L(u_o, (L\psi) \wedge f).$$

Inequality (6.9) is an immediate consequence of the maximum principle for parabolic operators, applicable here since (6.10) implies the following about the constant mapping $t \mapsto \phi(t) \equiv \phi$ from $[0,T]$ into $H^2(\Omega)$;

$$\begin{cases} \phi \in L^2(0,T;H^2(\Omega)) \quad \text{with} \quad \partial\phi/\partial t \in L^2(Q_T) \\ \phi(0) \leq u_T(0) \quad \text{a.e. in } \Omega, \quad \phi|_{\Sigma_T} \leq 0 \\ (\partial\phi/\partial t) + L\phi \leq (\partial u_T/\partial t) + Lu_T \quad \text{a.e. in } Q_T . \end{cases}$$

Thus, (6.7) follows from (6.8) and (6.9).

Because of (6.7), (6.6) implies

$$\|L\,Av(u_T)\|_{L^2(\Omega)} \leq \text{Constant independent of } T \in]0,+\infty[,$$

whence

$$\|Av\,(u_T)\|_{H^2(\Omega)} \leq \text{Constant independent of } T \in]0,+\infty[.$$

Thanks to the above, a sequence $T_n \to +\infty$ and an element u of $H_0^1(\Omega) \cap H^2(\Omega)$ exist, such that $u_n \equiv u_{T_n}$ satisfies

$$\text{(6.11)} \qquad Av(u_n) \to u \quad \text{in } H^2(\Omega).$$

The proof of the theorem will be completed by showing that

$$u = \sigma_L(\psi,f),$$

since the uniqueness of $\sigma_L(\psi,f)$ makes the use of the particular sequence $\{T_n\}_{n\in N}$ irrelevant.

It is easy to check that

$$u \in \phi_L(\psi,f).$$

Indeed, the inequality

$$u \leq \psi \quad \text{a.e. in } \Omega$$

follows from the inequality

$$Av(u_n) \leq \psi \quad \text{a.e. in } \Omega, \ \forall n \in N,$$

whereas the left-hand side inequality of (6.6), written for $T = T_n$, yields in the limit

$$Lu \leq f \quad \text{a.e. in } \Omega.$$

All that remains to be proven is that

$$\text{(6.12)} \qquad v \in \phi_L(\psi,f)$$

implies

(6.13) $\quad v \leq u \quad$ a.e. in Ω.

Let $\xi(t) \equiv (1+t)^{-1}$ and, for each $n \in N$ let z_n solve the (non-homogeneous) Cauchy-Dirichlet problem)

$$\begin{cases} z_n \in L^2(0,T_n);H^2(\Omega) \text{ with } (\partial z_n/\partial t) \in L^2(Q_{T_n}), \\ z_n(0) = u_o - \psi \quad \text{a.e. in } \Omega, \quad z_n|_{\Sigma_{T_n}} = \xi(u_o - \psi)|_{\Sigma_{T_n}} \\ (\partial z_n/\partial t) + Lz_n = 0 \quad \text{a.e. in } Q_{T_n} \quad . \end{cases}$$

Again by the maximum principle for parabolic operators we have

(6.14) $\quad z_n \leq 0 \quad$ a.e. in Q_{T_n}, $\forall\, n \in N$.

On the other hand, let χ solve the (non-homogeneous) Dirichlet problem

$$\chi \in H^2(\Omega), \quad \chi|_{\partial\Omega} = (u_o - \psi)|_{\partial\Omega}, \quad L\chi = (L(u_o - \psi)) \wedge 0$$
$$\text{a.e. in } \Omega$$

then the constant mapping $t \mapsto \chi(t) \equiv \chi$ satisfies, $\forall\, n \in N$,

$$\begin{cases} \chi \in L^2(0,T_n;H^2(\Omega)) \text{ with } (\partial\chi/\partial t) \in L^2(Q_{T_n}) \\ \chi(0) \leq z_n(0) \quad \text{a.e. in } \Omega, \quad \chi|_{\Sigma_{T_n}} \leq z_n|_{\Sigma_{T_n}} \\ (\partial\chi/\partial t) + L\chi \leq (\partial z_n/\partial t) + Lz_n \quad \text{a.e. in } Q_{T_n}, \end{cases}$$

hence

(6.15) $\quad z_n \geq \chi \quad$ a.e. in Q_{T_n}, $\forall\, n \in N$.

From (6.14) and (6.15) we deduce

(6.16) $\quad \|z_n(T_n)\|_{L^2(\Omega)} \leq$ Constant independent of $n \in N$;

since $Av(z_n)$ satisfies

$$\begin{cases} Av(z_n) \in H^2(\Omega), \quad Av(z_n)|_{\partial\Omega} = u_o - \psi|_{\partial\Omega}\,\big(\log(1+T_n)/T_n\big) \\ \big((z_n(T_n) - z_n(0))/T_n\big) + L\,Av(z_n) = 0 \quad \text{a.e. in } \Omega, \end{cases}$$

(6.16) yields

(6.17) $\quad \|Av(z_n)\|_{H^2(\Omega)} \to 0 \quad$ as $n \to +\infty$.

Now let

$$v_n \equiv v + z_n \quad \forall\, n \in N,$$

with v as in (6.12). Thus, each v_n satisfies

$$\begin{cases} v_n \in L^2(0,T_n;H^2(\Omega)) \text{ with } (\partial v_n/\partial t) \in L^2(Q_{T_n}) \\ v_n(0) \leq u_o \quad \text{a.e. in } \Omega, \quad v_n|_{\Sigma_{T_n}} \leq 0 \\ v_n \leq \psi, \quad (\partial v_n/\partial t) + Lv_n \leq f \quad \text{a.e. in } Q_{T_n} \end{cases}$$

and hence also [21]

(6.18) $\quad v_n \leq u_n \quad$ a.e. in Q_{T_n}, $\forall\, n \in N$.

Integrating (6.18) from 0 to T_n we obtain

$$v + Av(z_n) \leq Av(u_n) \quad \text{a.e. in } \Omega, \ \forall\, n \in N,$$

which yields (6.13) by a passage to the limit, once (6.11) and (6.17) are taken into account.

REFERENCES

1. A. Bensoussan and J. L. Lions, Nouvelle formulation de problèmes de contrôle impulsionnel et applications, *C. R. Acad. Sci. Paris A 276:* (1973) 1189.

2. A. Bensoussan and J. L. Lions, *Applications des inéquations variationnelles en contrôle stochastique*, Dunod, Paris, 1978.

3. A. Bensoussan and J. L. Lions, Book to appear.

4. M. Brezis and G. Stampacchia, Sur la régularité de la solution d'inéquations elliptiques, *Bull. Soc. Math. France 96:* (1968) 153.

5. L. Caffarelli and A. Friedman, Regularity of the solution of the quasi-variational inequality for the impulse control problem, *Comm. P.D.E. 3:* (1978) 745.

6. S. Campanato, Un risultato relativo ad equazioni ellittiche del secondo ordine di tipo non variazionale, *Ann. Sc. Norm. Sup. Pisa 21:* (1967) 701.

7. M. Chicco, Solvability of the Dirichlet problem in $H^{2,p}(\ )$ for a class of linear second order elliptic partial differential equations, *Boll. U.M.I. 4:* (1971) 374.

8. M. Chicco, Principo di massimo per soluzioni di equazioniellitiche del secondo ordine di tipo Cordes, *Ann. Mat. Pura Appl. 100:* (1974) 239.

9. H. O. Cordes, Zero order a priori estimates for solutions of elliptic differential equations, *Proc. Symp. Pure Math. 4:* (1961) 157.

10. B. Hanouzet and J. L. Joly, Convergence uniforme des itérés definissant la solution d'une inéquation quasi-variationnelle abstraite, *C. R. Acad. Sci. Paris A 286:* (1978) 735.

11. J. L. Joly and U. Mosco, A propos de l'existence et de la régularité des solutions de certaines inéquations quasi-variationnelles, to appear.

12. O. A. Ladyzenskaya, V. A. Solonnikov and N. N. Ural'Ceva, Linear and quasilinear equations of parabolic type, *Trabsl. Math. Mon. 23:* A.M.S., Providence, RI, 1968.

13. O. A. Ladyzenskaya and N. N. Ural'Cera, *Linear and quasi-linear elliptic equations,* Academic Press, New York, 1968.

14. J. L. Lions and G. Stampacchia, Variational inequalities, *Comm. Pure Appl. Math. 20:* (1967) 493.

15. M. E. Marina, Esistenza e unicità della soluzione di una disequazione variazionale associata a un operatore non coercivo, *Boll. U.M.I. 10:* (1974) 500.

16. U. Mosco, On some quasi-variational inequalities and implicit complementary problems of stochastic control theory, to appear.

17. H. H. Schaeffer, *Banach lattices and positive operators*, Springer-Verlag, Berlin, 1974.

18. G. Stampacchia, Formes bilinéaires coercitives sur les ensembles convexes, *C. R. Acad. Sci. Paris A 258:* (1964) 4413.

19. G. Talenti, Sopra una classe di equazioni ellittichea coefficienti misurabili, *Ann. Mat. Pura Appl. 69:* (1965) 285.

20. G. M. Troianiello, Unilateral Dirichlet problems of the non-variational type, to appear in *Ann. Math. Pura Appl.*

21. G. M. Troianiello, On a class of unilateral evolution problems, *Manuscripta Math. 29:* (1979) 353.

ON THE REGULARITY OF SOLUTIONS TO ELLIPTIC DIFFERENTIAL INEQUALITIES

J. FREHSE

Universität Bonn

C.E.R.E.M.A.D.E.
Université de Paris IX Dauphine

In this paper we discuss old and new techniques for proving the regularity of solutions

$$u \in L^{\infty}(\Omega) \cap H_0^{1,2}(\Omega)$$

of elliptic differential inequalities

(1) $$|Au| \leq K|\nabla u|^2 + K$$

in a bounded domain $\Omega \subset R^n$. Here K is some constant and A an elliptic operator,

$$Au = - \sum_{i,k=1}^{n} \partial_i (a_{ij}(x,u)\partial_k u),$$

whose coefficients satisfy

(2) $a_{ij}: \Omega \times R \to R$ *is Lipschitz continuous,*

(3) $$\sum_{i,k=1}^{n} a_{ij}(x,\mu)\xi_i\xi_k \geq c|\xi|^2$$

for all $x \in \Omega$, $\mu \in R$, $\xi = (\xi_1,\ldots,\xi_n) \in R^n$ with some constant $c > 0$ (ellipticity!).
Inequality (1) has to be understood in the sense of distributions in Ω. For the boundary $\partial\Omega$ of Ω we shall assume

(4') $\quad \partial\Omega \in C^{2+\gamma}$

with some $\gamma \in]0,1[$; or the following weaker condition which allows convex corners.

(4) *For every linear elliptic operator*

$$Bu = \sum_{i,k=1}^{n} b_{ik}(x)\partial_i\partial_k u$$

with Hölder continuous coefficients b_{ik} *and all* $p \in]2,\infty[$ *we have*

$$H^{2,p}(\Omega) \supset \{u \in H_0^{1,p}(\Omega) \mid \ Bu \in L^p(\Omega)\}.$$

The usual Lebesgue and Sobolev spaces are denoted by $L^p(\Omega)$, $H^{m,p}(\Omega)$, $H_0^{1,p}(\Omega)$, c.f. e.g. [15].

Under these hypotheses we have the following

THEOREM 1. Let $u \in L^\infty \cap H_0^{1,2}(\Omega)$ be a solution of (1) in a bounded domain $\Omega \subset R^n$ and let the conditions (2) - (4) on the data be satisfied. Then

$$Au \in L^\infty(\Omega) \quad \text{and} \quad u \in H^{2,p}(\Omega)$$

for all $p \in]1,\infty[$.

A similar theorem can be proved for elliptic systems provided that the solution $u = (u_1,\ldots,u_m)$ satisfies the stronger initial regularity $u \in C^\alpha \cap H_0^{1,2}(\Omega)$. Here $C^\alpha = C^\alpha(\Omega)$ denotes the space of Hölder continuous functions on Ω (with exponent $\alpha \in]0,1[$). To be precise, let $Au = ((Au)_1,\ldots,(Au)_m)$ and

$$(Au)_\nu = - \sum_{i,k=1}^{n} \sum_{\mu=1}^{m} \partial_i (a_{ik}^{\nu\mu}(x,u)\partial_k u_\mu).$$

The coefficients $a_{ik}^{\nu\mu}$ have to satisfy

(5) $a_{ik}^{\nu\mu}$: *is Lipschitz continuous*

(6) *For all* $\phi \in C_0^\infty(\Omega;R^m)$ *and* $u \in C^\alpha \cap H_0^{1,2}(\Omega;R^m)$ *we have*

$$\sum_{i,k=1}^{n} \sum_{\nu,\mu=1}^{m} \int_\Omega a_{ik}^{\nu\mu}(\cdot,u)\partial_k\phi_\mu \partial_i\phi_\nu dx \geq c \int_\Omega |\nabla\phi|^2 dx - \lambda_o \int_\Omega |\phi|^2 dx$$

with constants $c_o > 0$ *and* λ_o.

(Coerciveness / Legendre - Hadamard). Condition (4) now has to be understood for the system operator A.

THEOREM 2. Let $u \in C^\alpha(\bar{\Omega};R^m) \cap H_0^{1,2}(\Omega,R^m)$ satisfy the system of inequalities

$$|(Au)_\nu| \leq K|\nabla u|^2 + K, \qquad \nu = 1,\ldots,m$$

with some constant K in the sense of distributions in a bounded domain $\Omega \subset R^n$. Under the conditions (4) - (6) on the data we then have

$$Au \in L^\infty(\Omega;R^m) \quad \text{and} \quad u \in H^{2,p}(\Omega;R^m)$$

for all $p \in]1,\infty[$.

REMARKS.

(i) Theorem 1 and 2 are important for the regularity theory of quasi-linear elliptic equations and variational inequalities. If u satisfies an equation (or system) of the type

(5) $Au = F(\cdot,u,\nabla u)$

in the sense of distributions in Ω then u satisfies also an inequality of type (1) provided that

(6) $|F(x,u,\nabla u)| \leq K + K|\nabla u|^2$

and

(7) $F(x,\mu,\eta)$ *is measurable in* $x \in \Omega$ *and continuous in* $(u,\eta) \in R^m \times R^{mn}$.

Then Theorem 1 or 2 imply $H^{2,p}$ --regularity for u --under the required assumptions.

Another important application of Theorem 1 and 2 consists in the fact that solutions of variational inequalities with obstacles can be shown to satisfy an inequality of type (1). For example, consider the variational inequality

Find $u \in K = \{v \in H_0^{1,2}(\Omega) | \; u \leq \psi \text{ in } H^1\}$

such that $(Au, u-v) \leq (F(\cdot,u,\nabla u), u-v)$

for all $v \in K$.

Here $\psi\colon \Omega \to R$ is a given function with a certain degree of regularity, say $\psi \in H^{2,\infty}$.

Then one can prove the so called

"Lewy-Stampacchia-inequality" or "dual estimate"

$$0 \geq Au - F(x,u,\nabla u) \geq (0,[A\psi - F(x,u,\nabla u)]$$

c.f. [16], and also

$$0 \geq A(u) - F(x,u,\nabla u) \geq (0,[A\psi - F(x,\psi,\nabla\psi)]$$

c.f. [4], §5. Here $(f,g)(x) = \min(f(x),g(x))$.

(ii) The proofs of Theorem 1 and 2 which are discussed or presented in this paper yield also $H^{2,p}$ a priori estimates for u in terms of the data. Furthermore, interior regularity theorems can be obtained analogously.

METHODS OF PROVING THEOREM 1 AND 2

(A) The *two*-dimensional case was solved completely by E. Heinz [10] a long time ago, by strictly two-dimensional methods.

(B) Two important steps towards the solution of the regularity problem in consideration are contained in the work of Ladyzenskaya-Ural'zeva [12]. The first is the extension of the De Giorgi-Nash-Moser results to the case of equations $Au = F(\cdot,u,\nabla u)$ where F may have quadratic growth in ∇u. One obtains from their results that for every solution u of (1) we have

$$u \in C\,(\bar{\Omega}) \tag{8}$$

with some $\alpha \in]0,1[$ provided that the assumptions of Theorem 1 are satisfied. In fact, less regularity for $\partial\Omega$ and a_{ik} is necessary.

Inequality (1) implies

$$|(Au,\phi)| < K \int (|\nabla u|^2 + 1)\ |\phi| dx.$$

Choosing the test functions

$$\phi = \tau^2 \max \{u-k,0\} \qquad \text{and} \qquad \phi = \tau^2 \min \{u-k,0\}$$

with an appropriate localization function τ and constants k such that $\| \ \|_\infty < c/2K = \delta$ one obtains the integral relation

$$\int_{A_{k,\rho}} |\nabla u|^2 \tau^2 dx \leq K \int_{A_{k,\rho}} (u-k)^2 |\nabla\tau|^2 dx + K|A_{k,\rho}| \tag{9}$$

$$k \geq \max_{B_\rho} (u - \delta) \quad \text{resp.} \quad k \leq \min_{B_\rho} (u + \delta)$$

where $A_{k,\rho} = \{x \in B_\rho |\ u(x) \geq k \ \text{ resp. } \ u(x) \leq k\}$. B_ρ = ball of radius ρ.

From (9), the local Hölder continuity of u and the corresponding a priori estimate follows via the "classical" argument of De Giorgi and its generalization due to Ladyzenskaya-Ural'zeva [12] to the case where the lower order term $F(x,u,\nabla u)$ has quadratic growth in ∇u.

Other methods of reducing the latter case to De Giorgi's case can be found in [2], [3]. The Hölder continuity up to the boundary can be treated by variants of the above interior regularity methods c.f. [12], §2.7.

Once the C^α -estimates are established one can prove $H^{2,p}$ a priori estimate for the solution u of (1). A simple proof is given in the Appendix A. Under an additional coerciveness hypothesis for the nonlinear form

$$(Tu,u) = (Au,u) - F(\cdot,u,\nabla u),u)$$

one uses this to obtain an $H^{2,p}$ solution of the equation

$$(10) \qquad Au = F(\cdot,u,\nabla u), \qquad u \in H_0^{1,2} \cap H^{2,p}.$$

This procedure, however, does not imply a regularity theorem for *every* solution of (10). This problem was solved by Ladyzenskaya-Ural'zeva by proving the *local* uniqueness of solutions of (10). The procedure is as follows. Let $u \in L^\infty \cap H_0^1$ be a solution of (1). Then the theory of Ladyzenskaya-Ural'zeva yields $u \in C^\alpha(\bar{\Omega})$. Let $S_R = B_R \cap \Omega$ and consider the boundary value problem

$$v \in (u + H_0^{1,2}(S_R)) \cap L^\infty(S_R)$$

$$Av = F(\cdot,v,\nabla v) \quad \text{weakly in } S_R.$$

(Clearly, u itself is a solution of this problem.) Then a priori estimates for v can be obtained in the spaces $C^\beta(\bar{S}_R)$, $0 < \beta < \alpha$, $H_0^{1,2} \cap L^\infty(S_R)$, and $H_{loc}^{2,p}(S_R)$. Furthermore, one can derive that for every solution v the quantity

$$\underset{\partial S_R}{\operatorname{osc}} v = \underset{\partial S_R}{\max} v = \underset{\partial S_R}{\min} v$$

is small provided that R is small. This can be derived using the smallness of $\underset{\partial S_R}{\operatorname{osc}} u$ which is due to the fact that $u \in C^\infty(\bar{\Omega})$. From the a priori estimates one can derive the existence of a solution $v \in (u + H_0^{1,2}(S_R)) \cap H_{loc}^{2,p}(S_R)$. Then one can apply the theorem of local uniqueness which states that, for small R, the two solutions u and v must coincide, c.f. [12], §4.2. This yields that $u \in H^{2,p}_{loc}(\Omega)$; the $H^{2,p}$ regularity up to the boundary follows since $v \in H^{2,p}(\Omega \cap B_{R/2})$.

For the proof of the local uniqueness theorem it is essential that for small R the form $(Tz,w) = (Az,w) - (F(\cdot,z,\nabla z),w)$ is strictly monotone if z and w vary in a subset of functions $u + H^{1,2}_0(S_R)$ which have small oscillation and bounded C^β -norms.

A key lemma in this proof is Lemma 4.4 of §2, [12], which states that for $z \in C^\alpha \cap H^{1,2}(B_\rho)$, $\eta \in H^{1,2}_0(B_\rho)$

$$\int |\nabla z|^2 \eta^2 dx \leq K \rho^\alpha \int |\nabla \eta|^2 dx.$$

This lemma is used for estimating the term

$$(F(\cdot,z,\nabla z) - F(\cdot,w,\nabla w), z - w \leq \varepsilon_0 \int |\nabla z - \nabla w|^2 dx$$

(C) Ladyzenskaya-Ural'zeva do not cover the case of an *inequality* of type (1) and the case that $F(\cdot,u,\eta)$ is only continuous with respect to η. However, Fritz Tomi [18] showed by a simple trick that these cases can be reduced to those treated by Ladyzenskaya-Ural'zeva. He observed that any solution of (1) satisfies an equation

$$\text{(11)} \qquad Au = \sigma(x)K(|\nabla u|^2 + K)$$

where $\sigma = \sigma_u \in L^\infty(\Omega)$, $-1 \leq \sigma \leq 1$.

Equation (11) satisfies the hypotheses of the local uniqueness theorem in [12] and thus the problem of $H^{2,p}$ regularity of solutions of (1) was solved. Choosing $p > n$ one has $\nabla u \in L^\infty$ on account of Sobolev's theorem, and thus $Au \in L^\infty$.

(D) A way of proving Theorem 1 or 2 without using a theorem of local uniqueness consists in the method of confrontation due to Campanato [2] and later generalizations. By Tomi's trick it suffices to prove regularity for the equation (11). As it was worked out in [6], [8], [9] one can compare the solution u of (11) with the solution z of the boundary value problem

(12) *Find* $z \in u + H_0^{1,2}(S_R)$ *such that*

$$A_0 z = \sigma(x)K \quad \textit{in} \quad S_R = \Omega \cap B_R$$

where A_0 *is the "frozen operator"*

$$A_0 z = - \sum_{i,k=1}^{n} \partial_i (a_{ik}(x_0, \textstyle\int_{B_R} u\, dx)\, \partial_k z,$$

$\int_{B_R} = |B_R|^{-1} \int_{B_R}$. *Note that* $z \in H^{2,p}_{loc}(S_R)$.

Comparing the solution u of (1) and z of (2) one obtains (after several nontrivial steps) a Campanato condition

$$\int_{B_R} |\nabla u - (\overline{\nabla u})_R|^2 dx \leq KR^{n+2\alpha}, \qquad (\overline{\nabla u})_R = \int_{B_R} \nabla u\, dx.$$

From this we conclude $\nabla u \in C^{\alpha}$, $0 < \alpha < 1$ and thus $Au \in L^{\infty}$ from which the $H^{2,p}$ regularity follows.

(E) A new elegant way of proving Theorem 1 and 2 is the "interpolation method" which is worked out in the following. It consists of several steps (i) - (iv).

(i) From the theory of Ladyzenskaya-Ural'zeva [12] we obtain that the solution u of (1) is contained in $C^{\alpha}(\Omega)$ for some $\alpha \in]0,1[$. In the case of systems (Theorem 2), this is part of the hypothesis.

(ii) Inequality (1) implies

(13) $$(Au,\phi) \leq K \int |\nabla u|^2 |\phi| dx + K \int |\phi| dx$$

for all $\phi \in L^{\infty} \cap H_0^{1,2}(\Omega)$. We set

$$\phi = (u - \bar{u}_{2R})\tau^2$$

where τ is a Lipschitz continuous localization function such that

$$\tau = 1 \text{ on } B_R,\ \tau = 0 \text{ outside } B_{2R},\ |\nabla\tau| \le R^{-1}$$

and $\bar{u}_{2R}$ is the mean value of u taken over the set B_{2R}. If $B_{2R} \cap \Omega \neq \emptyset$ we set $\bar{u}_{2R} = 0$. Here B_R and B_{2R} are concentric balls of radius R and $2R$. With this test function we obtain "Cacciopoli's inequality"

$$(14) \qquad \int_{B_R} |\nabla u|^2 dx \le KR^{-2} \int_{B_{2R}} |u - \bar{u}_{2R}|^2 dx + K|B_{2R} \cap \Omega|$$

(Outside of Ω, we have extended u by 0).

The verification of (14) is routine; one has to use the ellipticity condition or the coerciveness; furthermore one has to take into account that

$$K \int_{B_{2R}} |\nabla u|^2 |u - \bar{u}_{2R}| \tau^2 dx \le \varepsilon \int |\nabla u|^2 \tau^2 dx \quad \text{for } R \le R\varepsilon$$

This follows since u is Hölder continuous.

From (14) we obtain via the inequality of Sobolev-Poincaré

$$(15) \qquad \int_{B_R} |\nabla u|^2 dx \le K\Big(\int_{B_{2R}} |\nabla u|^s dx\Big)^{2/s} + K|B_{2R} \cap \Omega| R^{-n}$$

with $s = 2n/(n+2) < 2$, $\int_M$ denoting the mean value taken over M. From (15) we obtain by Gehring's lemma that there exists a number $p > 2$ that

$$(16) \qquad \nabla u \in L^p(\Omega).$$

Gehring's trick [5] was first used in the theory of quasi conformal mappings and it was Meyers and Elcrat [13] who first applied this trick to nonlinear elliptic equations. Their ideas were refined in [6], [7], [9] and these methods are an important new tool in nonlinear elliptic analysis. The way how to apply Gehring's lemma in order to conclude (16) is explained in Appendix B.

(iii) In the following, we use a theorem of C. Miranda and L. Nirenberg on the interpolation between C^α and $H_0^{2,p}(\Omega)$, c.f. [14], [17]. They proved the following inequality

(17) $$||\nabla u||_q \leq K||\nabla^2 u||_r^\theta [u]_\alpha^{1-\theta} + K[u]_\alpha$$

where

$$1/q = (1/n) + \theta((1/r) - (2/n)) - (1 - \theta)(\alpha/n)$$

for all θ such that

$$(1 - \alpha)/(2 - \alpha) \leq \theta \leq 1.$$

We choose $\theta = 1/2$ and obtain

(18) $$1/q = (1/2r) - (\alpha/2n) < (1/2r)$$

Now, if $\nabla u \in L^p(\Omega)$, $p > 2$, we derive from (1) that

$$Au \in L^{p/2}(\Omega).$$

Since A can be considered as a linear elliptic operator with Hölder continuous coefficients we obtain from the theory of linear elliptic equations

$$\nabla^2 u \in L^{p/2}(\Omega)$$

and by (17) and (18)

(19) $$\nabla u \in L^q$$

where

(20) $$1/q = (1/p) - (\alpha/2n) < (1/p).$$

Thus we have obtained a higher L^q-exponent for ∇u. The statements (19) and (20) can be iterated ("bootstrap argument," "Münchhausen principle"): By (ii) we know already that

$$\nabla u \in L^{p(0)}(\Omega)$$

with

$$p(0) = 2 + 2\varepsilon > 2.$$

From (19) and (20) we then obtain

$$\nabla u \in L^{p(i)}(\Omega)$$

where

$$(1/p(i)) = (1/p(i-1)) - (\alpha/2n) = \dots = (1/p(0)) - (i\alpha/2n)$$

provided that $1/p(i) > 0$. For the largest index i which is admissible we have

$$(1/p(i)) - (\alpha/2n) < 0$$

that is

$$p(i) > 2n\alpha^{-1} > 2n.$$

It follows that $\nabla u \in L^p(\Omega)$, $p > 2n$, and from the differential inequality $\nabla^2 u \in L^s(\Omega)$, $s > n$. By Sobolev's lemma we conclude $\nabla u \in L^\infty(\Omega)$ and, using (1) again, that $Au \in L^\infty$. Hence $\nabla^2 u \in L^p(\Omega)$ for all $p < \infty$.

APPENDIX A. <u>The</u> $H^{2,p}$ <u>a priori estimate</u>

THEOREM A. Let $u \in H^{2,p}(\Omega) \cap H_0^{1,2}(\Omega) \cap L^\infty(\Omega)$, $p > 2$, be a solution of (1) and assume that the hypotheses of Theorem 1 or 2 hold. Then there is a constant C such that

$$u_{H^{2,p}(\Omega)} \leq C.$$

C depends on n, p, $|\Omega|$, $\partial\Omega$, $\|a_{ik}\|_{1,\infty}$, c, λ_0, K, $[u]_\alpha$, $\|u\|_{1,2}$, and can be chosen uniform for fixed Ω and n as p, $\|a_{ik}\|_{1,\infty}$, K, $[u]_\alpha$, $\|u\|_{1,2}$ vary in bounded sets and c, λ_0 are uniformly bounded from below by a positive constant.

Proof. By a well known interpolation theorem (which can be proved by partial integration) we conclude from the hypothesis $u \in H^{2,p} \cap L^\infty \cap H_0^{1,2}$ that

$$\nabla u \in L^{2p}(\Omega).$$

From the theory of *linear* elliptic differential operators (L^p-estimates) we obtain that

$$\text{(A1)} \qquad \int_{Q_R} |\nabla^2 u|^p dx \leq K \int_{Q_{2R}} |\nabla u|^{2p} + K_R.$$

Here $Q_R = Q'_R \cap \Omega$, $Q_{2R} = Q'_{2R} \cap \Omega$, and Q'_R, Q'_{2R} are concentric cubes with side length R and $2R$. By partial integration we obtain

$$\int_{Q_{2R}} |\nabla u|^{2p} dx \leq \int_{Q_{4R}} \tau^{2p} |\nabla u|^{2p} dx$$

$$\leq K \int_{Q_{4R}} |u - u_0| |\nabla^2 u| |\nabla u|^{2p-2} dx + K \int_{Q_{4R}} |\nabla\tau| \tau^{2p-1} |u - u_0| |\nabla u|^{2p-1} dx.$$

Here K is the usual generic constant and

$$u_0 \in [\min_{Q_{4R}} u, \max_{Q_{4R}} u] \quad \text{if } Q'_{4R} \cap \Omega = \emptyset$$

and

$$u_0 = 0 \quad \text{if } Q'_{4R} \cap \Omega \neq \emptyset.$$

Furthermore, τ is a Lipschitz continuous function such that $\tau = 1$ on Q_{2R}, $\tau = 0$ outside Q_{4R}, $|\nabla\tau| \leq KR^{-1}$.

Using Hölder's inequality we obtain

$$\int_{Q_{2R}} |\nabla u|^{2p} dx \leq \varepsilon \int_{Q_{4R}} |\nabla^2 u|^p dx + K_\varepsilon \|u - u_0\|_{L^\infty(Q_{4R})}^{p/(p-1)} \int_{Q_{4R}} |\nabla u|^{2p} dx.$$

If R has been chosen small enough, we have on account of the Hölder continuity of u that

$$K_\varepsilon \|u - u_0\|^{p/(p-1)}_{L^\infty(Q_{4R})} \leq \delta.$$

In fact, a uniform Hölder estimate for u is available via the theory of Ladyzenskaya-Ural'zeva (c.f. the discussion in the beginning of the paper).

Thus we obtain

$$\sup\{ \int_{Q_{2R}(z)} |\nabla u|^{2p} dx | \; z \in \Omega\}$$

$$\leq K_\varepsilon \sup\{ \int_{Q_R(z)} |\nabla^2 u|^p dx | \; z \in \Omega\}$$
$$+ K_\varepsilon \delta \sup\{ \int_{Q_R(z)} |\nabla u|^{2p} dx | \; z \in \Omega\}.$$

Note that $\sup \int_{Q_{4R}(z)} \leq K \sup \int_{Q_R(z)}$.

Using (A1) we arrive at the inequality

$$(1 - K_\varepsilon - K_\varepsilon \delta) \sup\{ \int_{Q_R(z)} |\nabla^2 u|^p dx | \; z \in \Omega\} \leq K_R$$

which proves the theorem.

We mention a recent paper of Amann-Crandal [1] where they prove the $H^{2,p}$ a priori estimate in the case $A = \Delta$ (Laplacean). They replace the C^α a priori estimate by a more elementary argument using the maximum principle. Also Tomi [18] is able to avoid the difficult C^α a priori estimates of Ladyzenskaya-Ural'zeva [12].

APPENDIX B. Application of Gehring's Lemma

LEMMA B. Let $\partial\Omega$ be Lipschitz and $u \in H^{1,2}(\Omega)$ satisfy

(B1) $$\int_{B_R} |\nabla u|^2 dx \leq K(\int_{B_{2R}} |\nabla u|^s dx)^{2/s} + K, \qquad s = 2n/(n+2)$$

for all balls $B_R \subset R^n$. Then there exists a number $p > 2$ such that $\nabla u \in L^p(\Omega)$.

Proof. Let $\chi(\Omega)$ be the characteristic function of Ω and $g = \max\{|\nabla u|^s, \chi(\Omega)\}$. Let Mh denote the maximal function of any real L^1-function h, i.e.

$$(Mh)(z) = \sup\{ \int_{B_R(z)} h\,dx \mid R > 0\}.$$

From inequality (B1) we derive

$$\text{(B2)} \qquad \int_{B_R(z)} |\nabla u|^2 dx \le K(Mg(z))^{2/s}.$$

(We have used the fact that $1 \le K(M\chi_\Omega(z))^{2/s}$; for $z \in \partial\Omega$ the hypothesis on $\partial\Omega$ is used.) From (B2) we obtain

$$\text{(B3)} \qquad (Mg^{2/s})(z) \le K(Mg(z))^{2/s}, \qquad 2/s > 2.$$

Now, Gehring's lemma [5] states that if inequality (B3) holds for a function $g \in L^2$ then $g \in L^p(\Omega)$ for some $p > 2$. The lemma follows.

APPENDIX C. $H^{2,p}$ Regularity in the General Quasi-linear Case

We consider the scalar differential inequality

$$\text{(C1)} \qquad |Au| \le K|\nabla u|^2 + K$$

with

$$Au = -\sum_{i=1}^{n} \partial_i a_i(x,u,\nabla u)$$

where the a_i satisfy the following conditions:

(C2) *The derivatives* $(\partial/\partial x_k)a_i(x,\mu,\eta) = a_{ix_k}(x,\mu,\eta)$,

$(\partial/\partial\mu)a_i(x,\mu,\eta) = a_{io}(x,\mu,\eta)$, $(\partial/\partial\eta_i)a_i(x,\mu,\eta) = a_{ik}(x,\mu,\eta)$,

$x \in R^N$, $\mu \in R^1$, $\eta \in R^n$, $i,k = 1,\ldots,n$,

are measurable in x *and continuous in* (μ,η).

(C3) *There is a constant* $K = K(C)$ *such that*

$$|a_{io}(x,\mu,\eta)| + |a_{ix_k}(x,\mu,\eta)| \leq K|\eta|^2 + K$$

$$|a_{ik}(x,\mu,\eta)| \leq K, \quad i,k = 1,\ldots,n,$$

for all $\eta \in R^n$, $\mu \in R$ *with* $|\mu| \leq C$.

(C4) *There is a constant* $\lambda_0 > 0$ *such that*

$$\sum_{i,k=1}^{n} \xi_i\xi_k a_{ik}(x,\mu,\eta) \geq \lambda_0 \cdot |\xi|^2$$

for all $\xi \in R^n$, $\mu \in R$, $\eta \in R^n$.

THEOREM C1. Let $u \in L^\infty \cap H_0^{1,2}(\Omega)$ be a solution of (C1) and assume the conditions (C2) - (C4) and (4') for the data. Then $u \in H^{2,p}(\Omega)$ for all $p \in]1,\infty[$.

Proof. By Tomi's trick, it suffices to study the equation

$$\text{(C5)} \qquad Au = \sigma(x)K(|\nabla u|^2 + 1), \qquad \sigma \in L^\infty, \quad -1 \leq \sigma \leq 1.$$

The theory of Ladyzenskaya-Ural'zeva yields $u \in C^\alpha$ and

$$\int_\Omega |\nabla u|^2 |x - x|^{2-n-2}\, dx \leq K \qquad \text{for all } z \in \Omega.$$

Since the solutions of (C5) are locally unique (in the sense discussed in the paper) it suffices to prove an $H^{2,p}$ a priori estimate for solutions z of

$$Ax = \sigma K|\nabla z|^2 + \sigma K, \quad z \in u + H_0^{1,2}(S_R), \quad S_R = B_R \cap \Omega$$

(More precisely, if $B_R \cap \Omega \neq \emptyset$, then $H^{2,p}_{loc}(S_R)$ estimates and $H^{2,p}$ estimates in a neighborhood of a smooth part of $\partial\Omega \cap B_R$ have to be established. We sketch here only the $H^{2,p}_{loc}$ estimates).

If $z \in H^{2,p}_{loc}$, $p \in]1,\infty[$ and $Az = \sigma(x)K(|\nabla z|^2 + 1)$ then

$$\sum_{i=1}^{n} (\partial_j \phi_i(x,z,\nabla z), \partial_i \phi) = -(\sigma K|\nabla z|^2 + \sigma K, \partial_j \phi), \quad \phi \in C_0^\infty(\Omega)$$

and

$$\text{(C7)} \qquad \sum_{i,k=1}^{n} (a_{ik}(\cdot,z,\nabla z)\partial_k \partial_j z, \partial_i \phi) \leq K \int_\Omega (|\nabla z|^2 + K)|\partial_j \phi| dx$$

$$j = 1,\ldots,n.$$

We may set $\phi = \tau^2 G \partial_j z$ where $\tau \in C_0^\infty(S_R)$ and G is the Greenfunction of the operator L defined by

$$Lv = -\sum_{i,k=1}^{n} \partial_i (a_{ik}(\cdot,z,\nabla z)\partial_k v).$$

We have

$$LG = \delta(\cdot - y), \quad G = G_y, \quad \langle\delta(\cdot - y), v\rangle = v(y).$$

Note that $G \in H^{1,1}(\Omega)$ and $z \in H^{2,p}$; thus the above choice of ϕ in (C7) is admissible.

The left hand side of (C7) can be estimated from below by

$$\text{(C8)} \qquad \tau^2(y)|\partial_j z(y)|^2 + \int |\nabla \partial_j z|^2 \tau^2 \; G \, dx$$

and lower order terms and terms which make the same difficulty as the right hand side of (C7).

The later term is estimated by using the fact that

$$\int |\nabla z|^2 G |x - y|^{-2} \, dx < \infty$$

which follows from the pointwise estimate

$$|G_y(x)| \leq K|x - y|^{2-n}, \qquad n \geq 3,$$

and the Hölder continuity of z. Thus, for $0 < \alpha' < \alpha$,

$$\int_{B_R} |\nabla z|^2 G|x-y|^{-2\alpha'}dx < \varepsilon^2$$

if B_R is small enough.

Furthermore, we use that

$$\int |\nabla G|^2 G^{-1}|x-y|^{2\alpha'}dx < \infty.$$

These remarks in mind, we estimate

$$\int |\nabla u|^2|\partial_j^2 u|G\tau^2 dx$$
$$\leq \varepsilon \int_\Omega |\partial_j^2 u|^2\tau^2 G\,dx + \varepsilon^{-1}\|\tau\partial_j u\|_\infty^2 \int_{B_R} |\nabla u|^2 G\,dx$$
$$\leq \varepsilon \int_\Omega |\partial_j^2 u|^2\tau^2 G\,dx + \varepsilon\|\tau\partial_j u\|_\infty^2$$

and

$$\int |\nabla u|^2|\partial_j u||\partial_j G|\tau^2 dx$$
$$\leq \varepsilon\|\tau\partial_j u\|_\infty^2 \int_{B_R} |\nabla G|^2 G^{-1}|x-y|^{2\alpha'}dx$$
$$+ K_\varepsilon \int |\nabla u|^4 G|x-y|^{-2\alpha'}\tau^2 dx$$
$$\leq \varepsilon\|\tau\partial_j u\|_\infty^2 + K_\varepsilon\|\tau\nabla u\|_\infty^2 \int_{B_R} |\nabla u|^2 G|x-y|^{-2\alpha'}dx$$
$$\leq \varepsilon\|\tau\partial_j u\|_\infty + \varepsilon\|\tau\nabla u\|_\infty^2 .$$

Thus we have arrived at the inequality

$$\int |\nabla u|^2|\partial_j u|dx \leq \varepsilon\|\tau\nabla u\|_\infty^2 + \varepsilon\int |\partial^2 u|^2 G\tau^2 dx.$$

From (C7) and (C8) and the above estimates (summing $j = 1,\ldots,n$) we obtain that

$$\|\tau\nabla u\|_\infty^2 + \int |\partial^2 u|^2 G\tau^2 dx$$

can be estimated by lower order terms.

Once having the above L_{loc} estimate for ∇u the right hand side of (C1) gives no difficulty anymore as far as the term $|\nabla u|^2$ is concerned. Now we can choose in (C7) the function

$$\phi = \tau^2 G(\partial_j u - c)$$

as test function and we obtain the Hölder continuity of ∇u via the hole-filling technique, c.f. e.g. [11], [3], §1.

The $H^{2,p}_{loc}$ estimate then follows from the linear theory of elliptic equations since $a_{ik}(x,u,\nabla u) \in C^{\alpha}$.

REFERENCES

1. H. Amann and J. M. Crandall, On some existence theorems for semi-linear elliptic equations, *Indiana Math. J. 27:* (1978) 779-90.

2. S. Campanato, Equazioni ellittiche del II ordine e spazi $\mathcal{L}^{2,\lambda}$, *Ann. Mat. pura e Appl. 69:* 321-382 (1965).

3. J. Frehse, On the smoothness of solutions of variational inequalities with obstacles, *Proc. Sem. Partial Diff. Equations*, Banach Center, Warszawa (1978).

4. J. Frehse and U. Mosco, Irregular obstacles and quasi-variational inequalities of stochastic impulse control, to appear.

5. F. W. Gehring, The L^p-integrability of the partial derivatives of a quasi-conformal mapping, *Acta Math. 130:* (1973) 265-277.

6. M. Giaquinta and E. Giusti, Non linear elliptic systems with quadratic growths, *Manuscripta Math. 24:* (1978) 323-349.

7. M. Giaquinta and G. Modica, Regularity results for some classes of higher order non linear elliptic systems, *J. für Reine u. Angewandte Math. 311/312:* (1979) 145-169.

8. M. Giaquinta, *Regularity results for weak solutions to variational equations and inequalities for non linear elliptic systems...*, Preprint 54, Heidelberg (1980).

9. M. Giaquinta, Remarks on the regularity of weak solutions to some variational inequalities, to appear in *Math. Z.*

10. E. Heinz, On certain non linear elliptic differential equations and univalent mappings, *J. Analyse Math. 5:* (1956-1957) 197-272.

11. S. Hildebrandt and K. O. Widman, On the Hölder continuity of weak solutions of quasi-linear elliptic systems of second order, *Ann. Sc. Norm. Sup. Pisa 1:* (1977) 145-178.

12. O. A. Ladyzenskaya and N. N. Ural'zeva, *Linear and quasi-linear elliptic equations,* New York: Academic Press (1968).

13. N. Meyers and A. Elcrat, Some results on regularity for solutions of non-linear elliptic systems and quasi-regular functions, *Duke Math. Journal 42:* (1975) 121-136.

14. C. Miranda, Su alcuni teoremi di inclusione, *Ann. Polon. Math. 16:* (1965) 305-315.

15. C. B. Morrey, Jr., *Multiple integrals in the calculus of variations,* Berlin: Springer (1966).

16. U. Mosco, Implicit variational problems and quasi-variational inequalities. From: L. Herausgeber and Waelbrock, Eds., *Non linear operators and the calculus of variations,* Lecture Notes in Mathematics, Springer.

17. L. Nirenberg, An extended interpolation inequality, *Ann. Sc. Norm. Sup. Pisa 20:* (1966) 733-737.

18. F. Tomi, Variationsprobleme vom Dirichlet-Typ mit einer Ungleichung als Nebenbedingung, *Math. Z. 128:* (1972) 43-74.

A PARTIAL DIFFERENTIAL EQUATION ARISING IN THE SEPARATION PRINCIPLE OF STOCHASTIC CONTROL

A. BENSOUSSAN

C.E.R.E.M.A.D.E.
Université de Paris IX Dauphine

INTRODUCTION

The separation principle of stochastic control obtained for the first time by Wonham [1] relies on the study of a quasi linear P.D.E. which presents the difficulty of having unbounded coefficients. This prevents the use of variational techniques. In A. Bensoussan and J. L. Lions [2] this equation (and the corresponding variational inequality) has been studied using a general abstract result of J. L. Lions [4] and also using probabilistic techniques. The objective of this paper is to give a simpler analytic proof. It uses the policy iteration technique and therefore cannot be extended to general quasi linear equations with unbounded coefficients.

I. SETTING OF THE PROBLEM

1.1 Assumptions - Notation

We consider a stochastic dynamic system governed by the equation

$$(1.1)\quad \begin{cases} dx = (F(t)x(t) + B(t;v(t))dt + G(t)dw(t) \\ x(0) = \xi \end{cases}$$

where

(1.2) F $n \times n$ matrix measurable and bounded

G $m \times n$ matrix measurable and bounded

(1.3) $w(t)$ standard Wiener process, with values in R^m,

(1.4) $B(t;v)$: $(0,\infty) \times U_{ad} \to R^n$, where U_{ad} is a compact convex subset of R^k, measurable (Lebesgue) with respect to t and continuous in v, $|B(t;v)| \leq C$

(1.5) ξ Gaussian random variable with values in R^n, independent from $w(\cdot)$, with mean x and covariance matrix P_0.

The variable ξ and the Wiener process $w(t)$ are defined with respect to an underlying probability space $(\Omega,\mathcal{A},P)$. We note that for any $v(t) \equiv v(t;\omega) \in L^2((0,T) \times \Omega;\ dt \otimes dP;\ R^k)$, with values in U_{ad}, we can solve (1.1) as follows:

(1.6) $x(t) = x_1(t) + \beta(t)$

where

(1.7) $dx_1/dt = F(t)x_1 + B(t;v(t) \qquad x_1(0) = 0$

and

(1.8) $d\beta = F(t)\beta dt + G(t)dw(t) \qquad \beta(0) = \xi.$

We next define an observation process as follows

(1.9) $dz = J(t)x(t)dt + R^{1/2}(t)db(t) \qquad z(0) = 0$

where

(1.10) $\begin{cases} H \text{ matrix } p \times n \text{ measurable and bounded in } t; \\ b(t) \text{ is a Wiener process, standard, in } R^p \\ \text{independent of } \xi, w; \\ R \text{ symmetric bounded, positive definite; } R^{-1}(t) \text{ bounded.} \end{cases}$

(1.11) $z(t) = z_1(t) + \gamma(t)$

where

(1.12) $dx_1/dt = H(t)x_1 \qquad z_1(0) = 0$

and

(1.13) $d\gamma = H(t)\beta dt + R^{1/2}(t)db(t) \qquad \gamma(0) = 0.$

We define by $\mathcal{F}^t$ the family of σ-algebras

(1.14) $\mathcal{F}^t = \sigma(\gamma(s), s \leq t).$

The process $v(t)$ is the control process. We write

(1.15) $L^2_{\mathcal{F}} = \{v \in L^2((0,T) \times \Omega; dt \otimes dP; R^k, v(t) \text{ is } \mathcal{F}^t \text{ measurable a.e. } t\}.$

For any $v \in L^2_{\mathcal{F}}$ (with values in U_{ad}), the process $z(t)$ is well defined and generates a family of σ-algebras (depending on v)

(1.16) $\mathcal{Z}^t_v = \sigma(z(s), s \leq t).$

It is easy to check that $\mathcal{Z}^t_v \subset \mathcal{F}^t \; \forall t.$
We define

(1.17) $\mathcal{U} = \{v \in L^2_{\mathcal{F}} |\ v(t)$ is $\mathcal{Z}^t_v$ measurable a.e. t, and

$v(t) \in U_{ad}$ a.e., a.s.$\}$.

The set $\mathcal{U}$ is not empty (it contains deterministic controls). It is not difficult to check that

(1.18) For any $v \in \mathcal{U}$, $\mathcal{Z}^t_v = \mathcal{F}^t \ \forall t$.

We call $\mathcal{U}$ the set of *admissible controls*. A process control which is admissible, is adapted to the family of σ-algebras generated by the observation process. Note that this observation process depends itself of the control. That is why an a priori class of controls had to be defined.

Let us now consider

(1.19) $\ell(x,v,t)$ continuous in v and (Lebesgue) measurable

with respect to x,t, $|\ell(x,v,t)| \leq C(1 + |x|^2 + |v|^2)$

(1.20) $h(x) \in W^{2,\infty}_{loc}$ $|h(x)| \leq C(1 + |x|^2)$.

We define

(1.21) $$J(v(\cdot)) = E[\int_0^T \ell(x(t),v(t),t)dt + h(x(T))].$$

The problem of stochastic control we are interested in is the following

(1.22) $$\inf_{v(\cdot)\in\mathcal{U}} J(v(\cdot)).$$

1.2 Kalman Filter

Let $v \in \mathcal{U}$, we consider the stochastic differential equation

(1.23) $$\begin{cases} d\hat{x} = F(t)\hat{x}dt + B(t;v(t))dt \\ \qquad\qquad + P(t)H^*(t)R^{-1}(t)(dz - H(t)\hat{x}(t)dt) \\ \hat{x}(0) = x \end{cases}$$

where $P(t)$ is the solution of the Riccati equation

(1.24) $$\begin{cases} dP/dt = FP + PF^* - PH^*R^{-1}HP + GG^* \\ P(0) = P_0. \end{cases}$$

Equation (1.23) defines in a unique way the process $\hat{x}(t)$. Moreover we have

(1.25) $\hat{x}(t)$ is $\mathcal{J}_v^t$ measurable $\forall t$.

Setting $\varepsilon = x - \hat{x}$, it follows from (1.1) and (1.23) that

(1.26) $$\begin{cases} d\varepsilon = (F = PH^*R^{-1}H)\varepsilon dt + Gdw - PH^*R^{-1/2}db \\ \varepsilon(0) = \tilde{\xi} = \xi - x. \end{cases}$$

Define $\hat{\beta}$ by

(1.27) $$\begin{cases} d\hat{\beta} = F(t)\hat{\beta}dt + PH^*R^{-1}(d\gamma - H\hat{\beta}dt) \\ \hat{\beta}(0) = x. \end{cases}$$

Then clearly

$$\varepsilon(t) = \beta(t) - \hat{\beta}(t) = x(t) - \hat{x}(t).$$

But $\hat{\beta}$ is the Kalman filter of β given the observation process γ (cf. for instance A. Bensoussan and J. L. Lions [2]), i.e.

(1.28) $\hat{\beta}(t) = E[\beta(t)|\mathcal{F}^t].$

Since

$$x(t) - \hat{x}(t) = \beta(t) - \hat{\beta}(t)$$

we deduce

$$\hat{x}(t) = E[E(t) \mid \mathcal{F}^t]$$

and from (1.18) it follows that

$$\hat{x}(t) = E[x(t) \mid \mathcal{F}_v^t]. \tag{1.29}$$

Therefore $\hat{x}(t)$ is the Kalman filter of the state $x(t)$ defined by (1.1), with respect to the observation process $z(t)$, for a given control $v(\cdot)$ in $\mathcal{U}$.

We note furthermore that $\varepsilon(t)$ defined by (1.26) is a Gaussian variable with mean 0 and covariance matrix $P(t)$.

1.3 Hamilton-Jacobi Bellman Equation

We define the operator

$$\tilde{A}(t) = -\sum_{i,j} \tilde{a}_{ij}(t)(\partial^2/\partial x_i \partial x_j) - \sum_{i,j} F_{ij}(t)x_j(\partial/\partial x_i) \tag{1.30}$$

where the matrix $\tilde{a}(t)$ is defined by

$$\tilde{a}(t) = (1/2)P(t)H^*(t)R^{-1}(t)H(t)P(t). \tag{1.31}$$

We note further

$$\tilde{\ell}(x,v,t) = E\ell(x + \varepsilon(t)v,t) \tag{1.32}$$

(1.33) $\tilde{h}(x) = Eh(x + \varepsilon(T))$.

We next introduce the Hamiltonian

(1.34) $$\tilde{H}(x,p,t) = \inf_{v \in U_{ad}} [\tilde{\ell}(x,v,t) + p.B(t;v)].$$

We will assume in the sequel the following

(1.35) $$\begin{cases} \tilde{a}(t) \geq \alpha I, \qquad \alpha > 0 \\ \tilde{a}_{ij}(t) \in C^0([0,T]). \end{cases}$$

From (1.35) it follows that the operator $\tilde{A}(t)$ is non degenerate. We note that it has unbounded coefficients. Our objective is to study in an adequate functional space the Cauchy problem

(1.36) $$\begin{cases} -(\partial u/\partial t) + \tilde{A}(t)u - \tilde{H}(x,Du,t) = 0 \\ u(x,T) = \tilde{h}(x). \end{cases}$$

II. STUDY OF THE H.J.B. EQUATION

2.1 Statement of the Main Result

Since we are considering a P.D.E. in the whole space, we will need the following Sobolev spaces with weights.

Define

(2.1) $\pi_s(x) = 1/(1 + |x|^2)^{s/2}$.

We set

$$(2.2)\qquad \begin{aligned} L^2_\pi &= \{z \mid \pi z \in L^2(R^n)\} \\ H^1_\pi &= \{z \in L^2_\pi,\ (\partial z/\partial x_i) \in L^2_\pi\}. \end{aligned}$$

The main result is the following

THEOREM 2.1. We assume (1.2), (1.4), (1.10), (1.19), (1.20), (1.35). Then there exists one and only one solution of (1.36) such that

$$(2.3)\qquad \begin{aligned} & u \in L^2(0,T;H^1_\pi) \\ & u \in L^p(0,T;W^{2,p}_{loc}(R^n)),\quad (\partial u/\partial t) \in L^p(0,T;L^p_{loc}(R^n)), \\ & \qquad\qquad \forall\ 2 \le p < \infty. \end{aligned}$$

2.2 Connection with the Stochastic Control Problem

Before proving Theorem 2.1 we give a hint on the connection between the Cauchy problem (1.36) and problem (1.22). For details see A. Bensoussan and J. L. Lions [2] or A. Bensoussan [1].

Define

$$(2.4)\qquad \tilde{L}(x,v,p,t) = \tilde{\ell}(x,v,t) + p.B(t;v).$$

For $\phi \in L^2(0,T;H^1_\pi)$ we write

$$(2.5)\qquad \tilde{L}_\phi(x,v,t) = \tilde{\ell}(x,v,t) + D\phi.B(t;v).$$

Then $\tilde{L}_\phi$ is continuous with respect to v and (Lebesgue) measurable in x,t. Since U_{ad} is compact, there exists a (Lebesgue) measurable function $v_\phi(x,t)$ such that

$$(2.6)\qquad \tilde{L}_\phi(x,v_\phi(x,t,),t) = \tilde{H}(x,D\phi,t) \quad \text{a.e.}$$

(see for instance Ekeland and Temam [3]).

Take $\phi = u$ (solution of 1.36) and define

$$(2.7) \qquad \hat{v}(x,t) = v_u(x,t).$$

More precisely we take a representative of $\hat{v}$ such that (2.6) holds $\forall\, x,t$. "If" we can solve the stochastic differential equation

$$(2.8) \qquad \begin{cases} d\hat{y} = F(t)\hat{y}dt + B(t;\hat{v}(\hat{y},t))dt + P(t)H^*(t)R^{-1}(t)(dz - H(t)\hat{y}(t)dt) \\ \hat{y}(0) = x, \end{cases}$$

and define $\hat{y}(t)$ adapted to $z(t)$, then the process

$$(2.9) \qquad \hat{v}(t) = \hat{v}(\hat{y}(t),t)$$

is an *optimal control*. At this stage the main difficulty consists in studying (2.8). One can do that under some drastic assumptions. It is also possible to generalize the class of admissible controls, in which case few additional assumptions are necessary. For details see A. Bensoussan and J. L. Lions [2] or A. Bensoussan [1]. We will not pursue this here, since this step is not necessary for the study of the Cauchy problem (1.36) itself. Let us just mention that when one can justify the writing (2.9), then the optimal control is obtained from a feedback, which is a deterministic function of the Kalman filter. This is called the separation principle in the sense that there is a separation between the operations of estimating and controlling.

2.3 Study of a Linear Equation

Our objective in this paragraph is to give a preliminary result for the proof of Theorem 2.1.

LEMMA 2.1. Let $f \in L^2(0,T;L^2_\pi)$, $\bar{u} \in L^2$. Let $a_{ij}(t)$ be a matrix such that

$$(2.10) \qquad a_{ij} = a_{ji}, \quad \sum_{i,j} a_{ij}\xi_j\xi_i \geq \alpha|\xi|^2 \ \forall\ \xi \in R^n, \quad \alpha > 0$$

and $g(x,t)\colon R^n \times (0,T) \to R^n$ such that

(2.11) g measurable and bounded.

Then there exists one and only one solution of

$$(2.12) \qquad \begin{cases} -(\partial u/\partial t) - \sum_{i,j} a_{ij}(t)(\partial^2 u/\partial x_i \partial x_j) - F(t)x.Du \\ \qquad\qquad - g(x,t).Du = f \\ u(x,T) = \bar{u}(x) \\ u \in L^2(0,T;H^1_\pi) . \end{cases}$$

Proof. Since $F(t)x$ is not a bounded function we cannot apply a variational inequality theory. In a first stage we will assume in addition that

(2.13) $f,\bar{u}$ are bounded.

Let us define

$$(2.14) \qquad P_k(x) = \begin{cases} x & \text{if } |x| \leq k \\ (x/|x|)k & \text{if } |x| \leq k \end{cases}$$

then there exists one and only one solution of

$$(2.15)\quad \begin{cases} -(\partial u_k/\partial t) - \sum_{i,j} a_{ij}(t)(\partial^2 u_k/\partial x_i \partial x_j) - F(t)P_k(x).Du_k \\ \qquad\qquad - g.Du_k = f \\ u_k(x,T) = \bar{u}(x) \\ u_k \in L^2(0,T;H^1_\pi) . \end{cases}$$

This is an application of the variational theory, since all coefficients are bounded. We have also, from the Maximum principle,

$$(2.16)\quad |u_k(x,t)| \le T\|f\|_{L^\infty} + \|\bar{u}\|_{L^\infty} .$$

Let us then obtain additional estimates. Multiply (2.15) by $u_k\pi_s^2$ and integrate over R^n. We obtain

$$(2.17)\quad -(1/2)(d/dt)|u_k(t)|^2$$

$$+ \sum \int_{R^n} a_{ij}(t)(\partial u_k/\partial x_j)\Big[(\partial u_k/\partial x_i)\pi^2$$

$$- 2\big(u_k\pi^2 s x_i/(1 + |x|^2)\big)\Big]dx$$

$$- \int_{R^n} F(t)P_k(x).Du_k u_k\pi^2 dx - \int_{R^n} g.Du_k u_k \pi^2 dx$$

$$= \int f u_k \pi^2 dx .$$

But from (2.16) we can assert that

$$|F(t)P_k(x)u_k|_{L^2_\pi} \le C .$$

Therefore we deduce from (2.17)

$$(1/2)|u_k(t)|^2_\pi + \alpha \int_t^T \|u_k(s)\|^2 ds \le C \int_t^T |u_k(s)|^2 ds + C$$

from which it follows that

$$(2.18) \qquad \|u_k\|_{L^2(0,T;H^1_\pi)} \leq C .$$

We can then consider a subsequence still denoted by u_k, such that

$$u_k \to u \quad \text{in} \quad L^2(0,T;H^1_\pi) \quad \text{weakly, and} \quad L^\infty(R^n \times (0,T))$$

weak star. Since $F(t)P_k(x) \to F(t)x$ in $L^2(0,T;L^2_\pi)$ strongly, we can pass to the limit in (2.15) as $k \to \infty$, and obtain a solution of (2.12). Let us now obtain an a priori estimate for such a solution. Multiplying (2.12) by $u\pi^2$ and integrating over R^n yields

$$(2.19) \qquad -(1/2)(d/dt)|u(t)|^2_\pi$$

$$+ \Sigma \int_{R^n} a_{ij}(t)(\partial u/\partial x_j)\Big[(\partial u/\partial x_i)\pi^2 + 2u\big(\pi^2 s x_i/(1+|x|^2)\big)\Big]dx$$

$$- \int_{R^n} F(t)x.Du\, u\, \pi^2 dx - \int_{R^n} g.Du\, u\, \pi^2 dx$$

$$= \int_{R^n} f\, u\, \pi^2 dx .$$

But

$$\int_{R^n} F(t)x.Du\, u\, \pi^2 dx = \int_{R^n} F(t)x \cdot (1/2)D(u^2)\pi^2 dx$$

$$= -(1/2) \int_{R^n} \mathrm{tr}\, F(t) u^2 \pi^2 dx$$

$$+ s \int_{R^n} u^2 \big(F(t)x.x/(1+|x|^2)\big)\pi^2 dx$$

$$\leq C|u(t)|^2_\pi .$$

From this and (2.19) we deduce

$$(1/2)|u(t)|_{\pi}^{2} + \alpha \int_{t}^{T} \|u(s)\|_{\pi}^{2} ds$$

$$\leq C[\int_{t}^{T} |u(s)|_{\pi}^{2} ds + |\bar{u}|_{\pi}^{2} + \int_{t}^{T} |f(s)|_{\pi}^{2} ds]$$

where the constant C depends on the L^{∞} bound of g, F, a_{ij}. Therefore we obtain

$$\|u\|_{L^{\infty}(0,T;L_{\pi}^{2})} \leq C(|\bar{u}|_{L^{2}(\pi)} + |f|_{L^{2}(0,T;L_{\pi}^{2})})$$

(2.20)

$$\|u\|_{L\ (0,T;H_{\pi}^{1})} \leq C(|\bar{u}|_{L^{2}(\pi)} + |f|_{L^{2}(0,T;L_{\pi}^{2})})$$

where C depends only on the L^{∞} bounds of g, F, a_{ij}, on α and the choice of s.

From this estimate and the linearity of (2.12) we deduce uniqueness, and also the fact that the existence and uniqueness result extends to data as in the statement of the Lemma. This completes the proof.

LEMMA 2.2. We assume

(2.21) $\quad a_{ij} \in C^{0}([0,T])$,

(2.22) $\quad f \in L^{p}(0,T;L_{loc}^{p}(R^{n}))$, $\bar{u} \in W_{loc}^{2,p}(R^{n})$, $2 \leq p < \infty$.

Then the solution of (2.12) satisfies

(2.23) $\quad u \in L^{p}(0,T;W_{loc}^{2,p}(R^{n}))$, $(\partial u/\partial t) \in L^{p}(0,T;L_{loc}^{p}(R^{n}))$.

Proof. Let $\phi \in \mathcal{D}(R^{n})$. We define $z = \phi u$. Then z is a solution of

$$(2.24)\qquad -(\partial z/\partial t) - \sum_{i,j} a_{ij}(\partial^2 z/\partial x_i \partial x_j)$$

$$= (f + F(t)x.Du + g.Du) - \sum_{i,j} a_{ij}(\partial^2\phi/\partial x_i \partial x_j)u$$

$$- 2 \sum_{i,j} a_{ij}(\partial\phi/\partial x_j)(\partial u/\partial x_i)$$

$$z(x,T) = \phi(x)\bar{u}(x).$$

Let $\mathcal{O}_\phi$ be a smooth bounded domain containing the support of ϕ. Hence z vanishes on $\partial\mathcal{O}_\phi$.

Note now that because of the term ϕ, there is no difference any more between the terms $F(t)x$ and g on the right hand side of (2.24). Since $u \in L^2(0,T;L^2(\mathcal{O}_\phi))$, $(\partial u/\partial x_i) \in L^2(0,T;L^2(\mathcal{O}_\phi))$, $z \in \mathcal{H}^{2,1}(\phi)$ where $Q_\phi = \mathcal{O}_\phi \times (0,T)$ from the standard regularity theorems. We then use a bootstrap argument to obtain the desired result.

2.4 Proof of Theorem 2.1

We use the policy iteration technique. Let u^0 be an arbitrary function satisfying the regularity properties stated in (2.3). For u^n known, define

$$v^n(x,t) = v_{u^n}(x,t) \qquad \text{(cf. (2.7))}.$$

Let u^{n+1} to be the solution of the linear equation

$$(2.25)\qquad -(\partial u^{n+1}/\partial t) - \sum_{i,j} \tilde{a}_{ij}(t)(\partial^2 u^{n+1}/\partial x_i \partial x_j) - F(t)x.Du^{n+1}$$

$$= \tilde{\ell}(x,v^n(x,t),t) + Du^{n+1}.B(t;v^n(x,t))$$

$$u^{n+1}(x,T) = h(x) .$$

From the existence and uniqueness of u^{n+1} we can apply Lemmas 2.1 and 2.2. Indeed $h \in H^1_{\pi_s}$ for s large enough and $h \in W^{2,p}_{loc}(R^n)$ for any p. Moreover setting

$$g^n(x,t) = B(t;v^n(x,t))$$

then from (1.4) it follows that g^n is bounded (the L^∞ bound being independent of n). Next setting

$$f^n(x,t) = \tilde{\ell}(x,v^n(x,t),t)$$

it follows from assumptions (1.19) that

$$|f^n(x,t)| \leq C(|x|^2 + 1)$$

the constant C being independent of n, hence f^n remains bounded in $L^2(0,T;L^2_{\pi_s})$ for s large enough and in $L^p(0,T;L^p_{loc}(R^n))$, for any p.

Therefore we can assert that the sequence u^n is well defined and satisfies

$$\text{(2.26)} \quad \begin{cases} \|u^n\|_{L^2(0,T;H^1_\pi)} \leq C \\ \|u^n\|_{W^{2,1,p}(Q)} \leq C_Q \quad \forall\, 2 \leq p < \infty \\ \text{and } Q = \mathcal{O} \times (0,T) \text{ where } \mathcal{O} \text{ is an arbitrary bounded} \\ \text{smooth domain of } R^n. \end{cases}$$

We next have

$$-(\partial/\partial t)(u^{n+1} - u^n) - \sum_{i,j} \tilde{a}_{ij}(t)(\partial^2/\partial x_i \partial x_j)(u^{n+1} - u^n)$$
$$- F(t)x.D(u^{n+1} - u^n)$$
$$\leq D(u^{n+1} - u^n).B(t,v^n)$$

$$(u^{n+1} - u^n)(x,T) = 0.$$

We multiply by $(u^{n+1} - u^n)^+ \pi^2$ and integrate over R^n. We obtain

$$|(u^{n+1} - u^n)^+(t)|_\pi^2 \leq \int_t^T (u^{n+1} - u^n)^+(s)|_\pi^2 ds$$

hence

$$(u^{n+1} - u^n)^+ = 0 .$$

Therefore the sequence u^n is decreasing. By the second estimate (2.6) we see that u^n is bounded in $\bar{Q}$, $\forall\, \mathcal{O}$, hence we may assert that

(2.27) $$u^n \downarrow u \text{ pointwise, in } L^2(0,T;H_\pi^1) \text{ weakly and } W^{2,1,p}(Q) \text{ weakly for any } p \geq 2,\ p < \infty .$$

By compactness we may also assert that

(2.28) $$Du^n \to Du \quad \text{in } L^p(Q) \text{ for any } p \geq 2,\ p < \infty,\ \forall\, Q = \mathcal{O} \times (0,T).$$

Let $v \in U_{ad}$ arbitrary, we have

$$-(\partial u^n/\partial t) - \sum_{i,j} \tilde{a}_{ij}(t)(\partial^2 u^n/\partial x_i \partial x_j) - F(t)x.Du^n - \tilde{\ell}(x,v,t) - Du^n.B(t,v)$$

$$\leq -(\partial u^n/\partial^t) - \sum_{i,j} \tilde{a}_{ij}(t)(\partial^2 u^n/\partial x_i \partial x_j) - F(t)x.Du^n - \tilde{\ell}(x,v^n,t) - Du^n.B(t,v^n)$$

and from (2.25)

$$= (\partial/\partial t)(u^{n+1} - u^n) + \sum_{i,j} \tilde{a}_{ij}(\partial^2/\partial x_i \partial x_j)(u^{n+1} - u^n)$$

$$+ F(t)x.D(u^{n+1} - u^n) + D(u^{n+1} - u^n).B(t,v^n)$$

$$\rightarrow 0 \text{ in } L^p(Q) \text{ weakly } \forall Q, \ \forall p,$$

by virtue of (2.27), (2.28).

Therefore letting $n \rightarrow \infty$ we obtain

$$-(\partial u/\partial t) - \sum_{i,j} \tilde{a}_{ij}(t)(\partial^2 u/\partial x_i \partial x_j) - F(t)x.Du - \tilde{\ell}(x,v,t) - Du.B(t,v)$$

$$\leq 0$$

and since v is arbitrary

$$-(\partial u/\partial t) - \sum_{i,j} \tilde{a}_{ij}(\partial^2 u/\partial x_i \partial x_j) - F(t)x.Du - \tilde{H}(x,Du,t) \leq 0.$$

A reverse inequality can be obtained by a similar argument, exchanging the roles of u and u^n. Let us now prove uniqueness. Let u^1, u^2 be two solutions and v^1, v^2 controls such that

$$v^i = v_{u^i}(x,t), \qquad i = 1,2 .$$

We have

$$-(\partial u^1/\partial t) - \sum_{i,j} \tilde{a}_{ij}(\partial^2 u^1/\partial x_i \partial x_j) - F(t)x.Du^1$$

$$= \tilde{\ell}(x,v^1,t) + Du^1.B(t,v^1)$$

$$\geq \tilde{H}(x,Du^2,t) + D(u^1 - u^2).B(t,v^1)$$

hence

$$(\partial/\partial t)(u^1 - u^2) + \sum_{i,j} \tilde{a}_{ij}(\partial^2/\partial x_i \partial x_j)(u^2 - u^1)$$
$$+ F(t)x.D(u^2 - u^1) + D(u^2 - u^1).B(t,v^1)$$
$$\geq 0$$

$$(u^2 - u^1)(x,T) = 0.$$

From this we deduce $u^2 - u^1 \geq 0$. By a symmetric argument we have $u^1 - u^2 \geq 0$. Hence $u^1 = u^2$.

REFERENCES

1. A. Bensoussan, *Stochastic control by functional analysis methods*, North Holland, to be published.

2. A. Bensoussan and J. L. Lions, *Applications des inéquations variationnelles en contrôle stochastique*, Dunod, Paris, 1978.

3. I. Ekeland, R. Temam, *Analyse convexe et problèmes variationnels*, Dunod, Paris, 1974.

4. J. L. Lions, *Equations différentielles opérationnelles et problèmes aux limites*, Springer Verlag, Berlin, 1961.

5. W. M. Wonham, On the separation theorem of stochastic control, *Siam J. Control 6 (2):* (1966).

ORDINARY DIFFERENTIAL EQUATIONS AND CONTROL

FROM CLASSICAL TO MODERN SIGNAL PROCESSING: SOME ASPECTS OF THE THEORY

P. BERNHARD

C.E.R.E.M.A.D.E.
Université de Paris IX Dauphine

This paper is derived from a talk that the author gave at the Mathematics colloquium of the C.E.R.E.M.A.D.E. in April of 1980. Its intent was to give a mathematical audience an idea of what signal processing is about, some of its problems, some of its answers, with, of course, emphasis on the modern theory.

It is an introductory paper, definitely not a survey, nor a synthetic presentation. In order to emphasize the origins of the problems, we adhered to a historic presentation. Proofs were kept to a minimum, and only given for some of the less classical results.

Born as an engineering problem with little mathematical influence, signal processing has been strongly revitalized by the two major technical challenges: Radar signals, which gave birth to Wiener's theory during World War II, and space tracking and telecommunications which, together with the advent of digital computers, made Kalman's theory flourish in the 1960's. More recent mathematical thinking has completely bridged the gap, melting the two approaches in a single theory. We elected to emphasize the history of this development, neglecting other important aspects like the Fast Fourier Transforms for instance, that we shall only hint at in the first section.

The first section opens with an explanation of the very meaning of the words "signal" and processing," with some early examples. This is an opportunity for a brief indication of sampling problems and FFT, that we shall not develop further in this paper. In Section 2 we introduce Wiener's formidable contribution, both in the way of formulating an estimation problem and in its solution. Section 3 is devoted to the classical Kalman filter. Sections 4 and 5 are devoted to the basic Gauss Markov identification theory of Faurre, which allows one to apply Kalman's filter to Wiener's problem, and to some comments which open the way to Section 7. Section 6 also makes a bridge with the ARMA representation of the statisticians, and Section 7 finishes to show the relationships between the two approaches to filtering.

1. Basic Problems

A signal is a time function $y(\cdot)$:

(1) $\quad t \to y(t)$

taking its values in R for the classical theory, or in R^m for the modern one. One should think of it as a measurement sequence of some kind, and processing it will mean deriving from it another time function $\hat{y}(t)$ which is supposed to be an improved measurement sequence, or represent the useful content of $y(t)$ in some sense.

A typical instance of the first case is the "filtering" or "smoothing" of an imperfect measurement sequence. We know the underlying physical process is smooth, and therefore the signal should be freed of its "shaky" appearance, which is due to measurement incertainties.

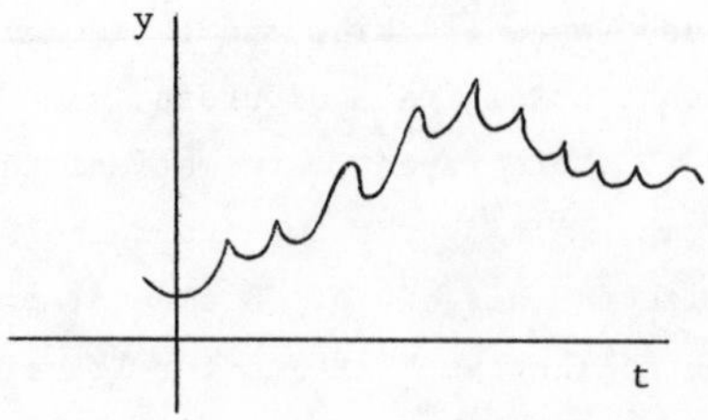

original signal

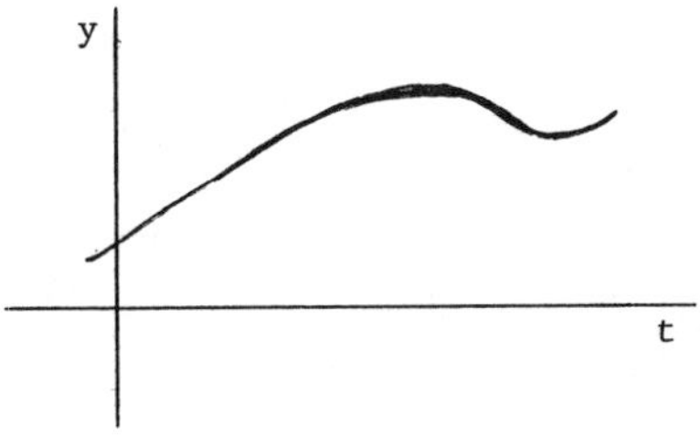

smoothed signal

An instance of the second problem is demodulating. The signal received may be perturbed sinusoid, with, say, varying amplitude. The useful signal is the amplitude.

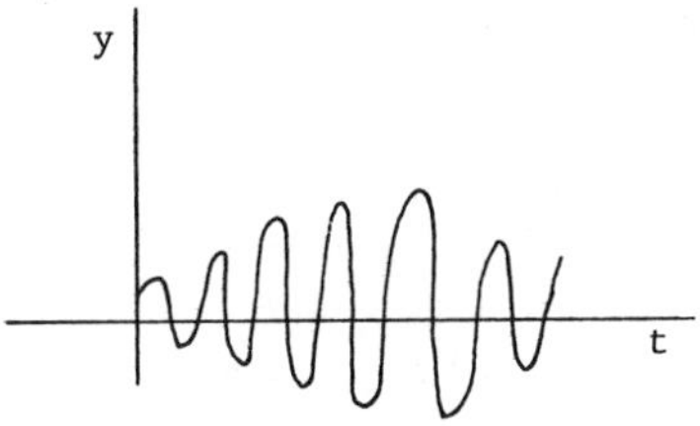

original signal

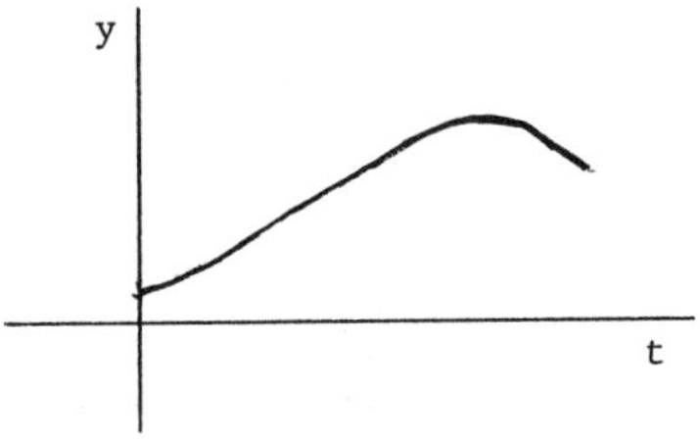

demodulated signal

Of course, both problems may be superimposed.

Other problems may involve attracting a known function with some unknown coefficients (a sinusoid with known frequency but unknown amplitude, say) from a mixture of parasite signals (other sinusoids with different frequencies for instance), deciding whether one known signal is present or not in a "noisy" signal (detection), etc.

One additional feature of signal processing is that we use both continuous and sampled (or discrete) signals, that is to say, the time t in (1) may range over R or over Z. This is made compulsory by the advent of digital processing and the use of computers, since these machines sample the signals at, possibly, very high, but yet finite, frequencies. The problem of the relationships between a continuous signal and its sampled version, the recovery of the first from the second, and the relation between their Laplace and "z" transforms (see below) is one of great interest, but will not be examined here.

We shall, in most of the sequel, prefer the discrete theory, (although the original Wiener theory was the continuous one) because it will enable us to avoid the technicalities of continuous diffusions and Ito calculus. Yet, all the important facts are there. The only drawback is that some formulas are slightly more complicated.

In the smoothing problem, it is clear that the perturbations we wish to remove are high frequencies as compared to the useful signal. This consideration, together with the fact that all this theory was related to electric circuits, therefore to linear transformations, whose eigen vectors are the exponentials, led to the use of frequency representations of the signal, i.e. Laplace transforms in the continuous case, and "z transforms" in the discrete case. These are defined in the following way: Let $\{y_k\}_{k\varepsilon Z}$ be a sequence of numbers, its z transform is

$$Y(z) = \sum_{k=-\infty}^{+\infty} y_k \, z^{-k}$$

If we decide that we shall always work with signals that are null for $k < k_0$ (usually zero), and bounded, then the above series has a finite radius of convergence. Its value for $z = e^{i\omega}$ is the discrete Fourier transform of the series. Very profound algebraic and number theoretic facts have led to extremely efficient ways of computing these Fourier Transforms. They are known as Fast Fourier Transforms or FFT algorithms.

A linear transformation acting on $y(\cdot)$, or *filter*, will be given by a convolution

$$\hat{y} = h * y,$$

where h is called the *impulse response* of the filter, or in Laplace transforms

$$\hat{Y}(z) = \mathcal{H}(z)\, Y(z),$$

and $\mathcal{H}(z)$ is known as the *transfer function* of the filter.

Therefore, the filter we would like to realize for the smoothing problem could be an ideal "low pass" filter, whose transfer function $\mathcal{H}(i\omega)$ in the continuous case, $\mathcal{H}(e^{i\omega})$ in the discrete case, would be equal to one for $|\omega| < \omega_0$, and to zero for $|\omega| > \omega_0$, for some well chosen ω_0.

For reasons explained by linear systems theory, and now well understood, only polynomial transfer functions can be obtained from electrical circuits, or from finite linear calculations with finite memory in the discrete case. Therefore, how to approximate such filters with realizable ones, what better shape given it to separate various signals, how to realize "band pass" and other related filters, are problems of interest to signal processing, and have been studied in depth.

The last feature of signal processing we must introduce is *causality*. If the measurement sequence (or the modulated signal) $y(t)$ is to be used for control purposes, we need the estimate $\hat{y}(t)$ almost simultaneously with the knowledge of $y(t)$, what is called in "real time." As a consequence, the filter processing y has to be causal, i.e. $\hat{y}(t)$ is a function of only past values of y:

$$\hat{y}(t) = f(\{y(s) \mid s < t\}) .$$

As a consequence, the impulse response of the filter must be null for negative t, and the transfer function must be holomorphic in the half plane $Re(z) > 0$ for the continuous case, or outside the unit disk in the discrete case.

2. Wiener Filter

The problem we address now is that of the first example of our previous section, and specifically with the requirement of real time processing, i.e. we want a causal filter.

The early theory was mainly qualitative. One was looking for a "good" filter, without precisely quantifying the problem. The first step in going beyond these methods is casting the problem into a framework within which "optimal" filtering has a meaning. This is achieved by using the device of bayesian estimation. One first assumes that the "true" signal, the one to be recoved, $z(\cdot)$, is a particular realization of an underlying stochastic process. The signal received, $y(\cdot)$ is another stochastic process, and some data is available about their joint distribution. Then the optimal filter is the one that gives:

$$\hat{z}(t) = E(z(t) \mid y(s),\ s < t).$$

The solution given by Wiener assumes that $y(\cdot)$ and $z(\cdot)$ are gaussian, stationary and ergodic. The necessary information is the autocorrelation function of $y(\cdot)$ (which can be measured since the process is assumed ergodic), $\Lambda_{yy}(t)$, and the cross-correlation function of $y(\cdot)$ and $z(\cdot)$: $\Lambda_{yz}(t)$.

Typically, we might have a model of the form:

$$(2) \qquad y(t) = z(t) + w(t),$$

where $w(\cdot)$ is a white sequence (a "white noise" in the continuous case), uncorrelated to $z(\cdot)$. Then, $\Lambda_{yy} = \Lambda_{zz} + R$, where R is the covariance of each random variable $w(t)$. (The spectral density in the continuous case.)

It is easy to see that y must be uncorrelated to $z - \hat{z}$, and as a consequence, the impulse response $h(t)$ of the optimal filter must satisfy

$$\Lambda_{zy}(t) = \sum_{k=0}^{\infty} h(k)\, \Lambda_{yy}(t - k).$$

We transform this relation with the z transform (there should be no confusion between the signal $z(\cdot)$ and the variable z of the z transform). We call spectrum the z transform of an autocorrelation function. Let them be $S_{zy}(z)$ and $S_{yy}(z)$. It comes

(3) $S_{zy}(z) = \mathcal{H}(z)\ S_{yy}(z).$

However, the solution is not $S_{zy}(z)\ S_{yy}^{-1}(z)$. As a matter of fact, this filter would not be causal, for with every zero of S_{yy} (a pole of S_{yy}^{-1}) inside the unit disk (the left half plane for the continuous case) the inverse (the opposite) is a zero (hence a pole of $S_{zy}S_{yy}^{-1}$).

The solution is obtained assuming that both spectrums are rational, it involves finding a strong factorization of $S_{yy}(z)$:(*)

(4) $S_{yy}(z) = W(z)\ W'(1/z),$

where W and its inverse are holomorphic outside the unit disk. (It is often denoted S_{yy}^{+}). It also uses the "causal part" of a rational matrix, denoted $\{\ \}_+$, which is obtained by keeping the causal terms of its expansion in simple factors. The transfer function of the optimal filter is given by:

(5) $\mathcal{H}(z) = \{S_{zy}(z)W'^{-1}(1/z)\}_+W^{-1}(z).$

In the scalar case, finding the strong factorization involves finding the roots of two polynomials, the numerator and the denominator of S_{yy}. However, in the vectorial case, S_{yy} is a square matrix, and performing the strong factorization is estremely difficult.

Finally, one must point out that once the optimal transfer function is known, the engineer is still faced with the problem of finding the electrical circuit (in the continuous case) that realizes it, a problem which gave birth to realization theory, one of the central parts of modern systems theory. In the discrete case, this translates into finding simple recursive calculations to perform on $y(t)$ to construct $\hat{z}(t)$ in "real time," with finite memory. (While programming the impulse response formula would require one to memorize all past values of y, i.e. to have a potentially infinite memory.)

(*) W' stands for W transposed.

3. Kalman Filter

To address the realization problem, Kalman starts with a different estimation problem, where the stochastic process to be estimated is given by a difference (or differential) equation, or *state-space* representation, of the form

$$(6) \qquad x(t+1) = Fx(t) + v(t), \qquad x(t) \in R^n.$$

The process $x(\cdot)$ is to be recovered from the observation of a set of linear combinations:

$$(7) \qquad y(t) = Hx(t) + w(t), \qquad y(t) \in R^p.$$

Here, $(v(\cdot), w(\cdot))$ is a white gaussian sequence of noises, with

$$E\left(\begin{pmatrix} v(t) \\ w(t) \end{pmatrix}, (v'(t), w'(t))\right) = \begin{pmatrix} Q & S \\ S' & R \end{pmatrix}$$

F, H, Q, S and R are known matrices of appropriate type. If $x(\cdot)$ and $y(\cdot)$ are to be stationary, these matrices must be constant, and F stable. However, it is a definite superiority of the new theory that it remains unchanged if they are time varying.

The change in the a priori model of the stochastic processes is the fundamental step taken toward the modern theory.

In the discrete case, it is quite simple to show that the optimal estimate

$$\hat{x}(t) = E(x(t) \mid y(t-1), y(t-2), \ldots)$$

is given by Kalman's filter

$$(8) \qquad \begin{aligned} \hat{x}(t+1) &= F\hat{x}(t) + K(t)(y(t) - H\hat{x}(t)), \\ \hat{x}(t_0) &= \hat{x}_0, \end{aligned}$$

where $\hat{x}_0$ is the a priori estimate of $x(t_0)$ (in the nonstationary case), with

$$(9) \qquad K(t) = (F\Sigma(t)H' + S)(H\Sigma(t)H' + R)^{-1}$$

$$(10) \qquad \Sigma(t+1) = F\Sigma(t)F' - (F\Sigma(t)H' + S)(H\Sigma H' + R^{-1}(H\Sigma(t)F' + S') + Q,$$

$$\Sigma(t_0) = \Sigma_0 \; .$$

Here, $\Sigma(t)$ is the covariance of $x(t) - \hat{x}(t)$, Σ_0 is the covariance of the a priori estimate (for the nonstationary case).

The continuous case is more difficult to establish rigorously, but leads to the same equation (8), and similar equations replacing (9) and (10). This last one is the celebrated Riccati equation, dual from that of control theory.

It can be proved that if the data are constant, $\Sigma(t)$ has a limit Σ as t goes to infinity, and therefore the filter (8) becomes stationary. And of course, one should notice that it is given in a finite memory form. The only things that must be memorized at each step are the last estimate $\hat{x}$, and, if it is time varying, its covariance Σ.

Since this filter is optimal, its stationary version cannot be anything else than a realization of the Wiener filter for x knowing y. However, we should emphasize that at this point, if that theory is more general in that it applies to nonstationary problems, it is more restricted in that it applies only to Markov processes of the type (6), and a very complete knowledge of this model has to be available. (In fact, every Markov process can be put in the form (6).)

4. Gauss Markov Realization

The next step to be taken is to use a Kalman-like theory to solve a problem in a form similar to Wiener's. To achieve this, we first consider a model of the form (6), but the signal to be recovered is now

(11) $z(t) = Hx(t),$

so that (7) has the same form as (2).

We have seen that Wiener's solution applies only to processes with a rational spectrum. Now, $(y(\cdot),z(\cdot))$ is such a process. As a matter of fact, let

(12) $\mathcal{H}(z) = H(zI - F)^{-1}$

be the transfer function of the system $v(\cdot) \to z(\cdot)$, by the classical theory of stochastic process, we have

(13) $$\begin{pmatrix} S_{zz}(z) & S_{zy}(z) \\ S_{yz}(z) & S_{yy}(z) \end{pmatrix} = \begin{pmatrix} \mathcal{H}(z) & 0 \\ \mathcal{H}(z) & I \end{pmatrix} \begin{pmatrix} Q & S \\ S' & R \end{pmatrix} \begin{pmatrix} \mathcal{H}'(1/z) & \mathcal{H}'(1/z) \\ 0 & I \end{pmatrix}.$$

Moreover, it is a fundamental result of realization theory that every proper rational spectrum can be represented as $S_{yy}(z)$ above, i.e. as the spectrum of a process of the form (6), (7).

Therefore, restricting oneself to this class of processes is equivalent to the restriction that the spectrum be rational. Of course, there is no need to assume that there actually exist a process $x(\cdot)$ that physically gives rise to $y(\cdot)$. Equations (6) and (7) may be considered as a mere mathematical model of the process $y(\cdot)$.

The question at this point is: is it possible, from the observation of the process $y(\cdot)$ to reconstruct (we say identify) the five matrices H, F, Q, S and R?

The answer to this question requires that we further analyze the properties of the stationary model (6), (7). It is easy to show that the covariance of $x(t)$ is the matrix P solution of the Lyapunov equation:

(14) $P - FPF' = Q,$

and that if we let

(15) $G = FPH' + S,$

The autocorrelation function of $y(\cdot)$ is given by

(16) $\Lambda(0) = HPH' + R,$

(17) $\Lambda(t) = HF^{t-1}G, \quad \forall\ t \in N.$

Formula (17) shows that the sequence $\Lambda(t)$, $t \in N$ is identical to the impulse response of the linear system (H,F,G). Therefore, using standard algorithms, like HO's or Riessanen's, it is possible to compute H, F and G from the knowledge of the autocorrelation, which can be deduced from measurements.

The triple of matrices of minimal dimension (for F) is unique, up to a change of basis on x, a very unessential non uniqueness.

Therefore the problem left is to find a positive semidefinite (or positive definite) matrix P such that the matrices Q, S and R deduced from H, F, G and P by equation (14), (15) and (6) satisfy

(18) $\begin{pmatrix} Q & S \\ S' & R \end{pmatrix} > 0.$

Unfortunately, the solution of this problem is usually non unique, and, of course, every solution gives rise to a model of the form (6), (7) for $y(\cdot)$. This non unicity is very essential. It says that there are several ways of distributing the noise intensity between dynamics and measurement, that yield the same statistics for the resulting dynamic process. We call them *equivalent* representations.

However, these representations can nevertheless be used unambiguously to filter $y(\cdot)$ thanks to the following fact, of which we shall give a very simple proof:

THEOREM. All equivalent representations of the same process $y(\cdot)$ have the same filtered estimate.

Proof. For a given representation, let

$$\hat{y} = H\hat{x} \tag{19}$$

be the corresponding estimate of y. Here $\hat{x}$ is the Kalman filter estimate. In equations (9) and (10), substitute the formulas in P, G and Λ_0 obtained from (14), (15) and (10) to Q, S and R. It turns out that the two equations can now be written in terms of

$$P_* = P - \Sigma \tag{20}$$

and read

$$K = (G - FP_*H')(\Lambda_0 - HP_*H')^{-1}, \tag{21}$$

$$P_* = FP_*F' + (G - FP_*H')(\Lambda_0 - HP_*H')^{-1}(G' - HP_*F') \tag{22}$$

(we have used the stationary version of (16)). It is known that (16) has several solutions, and the one that gives the filter is maximal for the ordering of positive definite matrices. Therefore, according to equation (20), (22) must have several solutions, the right one being minimal.

As a consequence, P_*, and thus K, are independent of the choice of P satisfying (18), and only depend upon H, F, G and Λ_0. The theorem is proved.

5. Faurre's Theorem

A complete understanding of Gauss Markov identification is mainly due to Faurre, and we shall give here an idea of that theory. A detailed modern account appears in [6].

Faurre has proved the following result:

THEOREM. The set $\mathcal{P}$ of matrices P that solve (14) to (18) is convex, the solutions of equation (22) are the extremal points of this set P_* is its minimal point, and yields the Kalman filter as realization.

Moreover, Faurre gave a rather explicit parametrization of the set $\mathcal{P}$, which is also shown to have a maximal element P^*, the maximal solution of (22), which corresponds to the antifilter, giving $E(y(t))| \ y(s), \ s > t)$.

The first thing to point out is that the Kalman filter is itself a representation of the form (6),(7). As a matter of fact, it is easy to see the *innovation process* $\eta(\cdot)$ given by

$$\eta(t) = y(t) - H\hat{x}(t)$$

is white, with covariance $\Lambda_0 - HP_*H'$, and (6) and (7) may be written

$$\hat{x}(t+1) = F\hat{x}(t) + K\eta(t),$$

$$y(t) = H\hat{x}(t) + \eta(t).$$

The particular fact is that in this representation, $v = Kw$, and therefore the matrix (18) is only positive semidefinite.

Conversely, if one has a realization having this property for some K, i.e. semidefinite, one has exactly

$$x(t+1) = Fx(t) + K(y(t) - Hx(t)),$$

so that

$$x(t) = E(x(t)/ \ y(t-1),\ldots)$$

(if $F - KH$ is stable), x is exactly known from the knowledge of y, and is therefore its own conditional expectation.

Now, let for a moment P_* be any positive semidefinite (or positive definite) matrix, and Q_*, S_*, R_* be the corresponding Q, S and R obtained through formulas (14), (15) and (16) (with H, F, G and Λ_0 given). Equation (22) can also be written

$$Q_* - S_* R_*^{-1} S_*' = 0$$

which is precisely the condition for (18) to be positive semidefinite (and not positive definite). It yields

$$V_* = K w_*$$

and

$$K = S_* R_*^{-1},$$

i.e. formula (21). Therefore the corresponding representation is the filter, and P_* is indeed the covariance of the corresponding state $x(t)$.

Now, for any other realization, with state $x(\cdot)$, let us write

$$x(t) = \hat{x}(t) + \tilde{x}(t).$$

$\tilde{x}$ is the error of estimation. It is uncorrelation to $\hat{x}$. Therefore, if we call Σ its covariance, we have

$$P = P_* + \Sigma$$

which is formula (20). But since Σ is necessarily positive, we see that P_* is minimal among all possible realizations.

6. ARMA Representation

Before we leave the time domain to go back to Wiener's formula, we want to show how the previous theory also allows one to find a multivariable ARMA (Auto Regressive Moving Average) model for $y(\cdot)$, a model widely used in statics and econometry.

The point is that one knows how to go from an internal realization such as the filter to an ARMA form. We shall show a simple way (though not minimal in a sense we shall precise) to do so, due to Alengrin and Favier. It is known that, via a change of basis on x (and on the output if necessary, this is unimportant here), one can put the pair (H,F) of (6), (7) in the Brunovski canonical form which

$$F = J + CH,$$

where J is a very simple matrix, which is nilpotent

$$J^k = 0$$

for some k (the largest observation invariant of the pair H,F).

Therefore the filter may be written

$$\hat{x}(t+1) = J\hat{x}(t) + Cy(t) + (K - C)\eta(t),$$

$$y(t) = H\hat{x}(t) + \eta(t).$$

We introduce the impulse response of this system. Let $A(0) = B(0) = I$, and

$$A(t) = HJ^{t-1}C, \qquad B(t) = HJ^{t-1}(K - C), \qquad t > 0.$$

The sequences $A(\cdot)$ and $B(\cdot)$ are null for $t > k$. We therefore have

$$y(t) = \sum_{\tau=1}^{k} A(\tau)y(t-\tau) + \sum_{\tau=0}^{k} B(\tau)\eta(t-\tau),$$

which is the ARMA model we were looking for. In z transform it becomes

$$\bar{A}(z)Y(z) = \bar{B}(z)\bar{H}(z),$$

where $Y(z)$ and $\bar{H}(z)$ are the z transforms of $y(\cdot)$ and $\eta(\cdot)$, and

$$\bar{A}(z) = I - \sum_{\tau=0}^{k} z^{k-\tau}A(\tau), \qquad \bar{B}(z) = \sum_{\tau=0}^{k} z^{k-\tau}B(\tau).$$

However, it can be shown that if k is indeed the smallest possible degree for such a representation, there usually remain a non unimodular common factor in $\bar{A}(z)$ and $\bar{B}(z)$, that can be removed to get a "minimal" form. (This factor has the form $\mathrm{diag}(1,z^{k_2},\ldots,z^{k_p})$).

7. Riccati Equation and Spectral Factorization

We turn back to the spectrum of y:

$$S_{yy}(z) = \sum_{t=-\infty}^{\infty} \Lambda(t)z^{-t}, \tag{23}$$

and we write it

$$S(z) = S_{yy}(z) = S_0(z) + \Lambda_0 + S_0'(1/z), \tag{24}$$

where

$$S_0(z) = \sum_{t=1}^{\infty} \Lambda(t)z^{-t}.$$

According to (17), we also have

$$S_0(z) = H(zI - F)^{-1}G. \tag{25}$$

Equation (13) also gives

$$(26) \qquad S(z) = [H(zI - F)^{-1} \quad I] \begin{bmatrix} Q & S \\ S' & R \end{bmatrix} \begin{vmatrix} ((1/z)I - F')^{-1} \\ I \end{vmatrix}$$

(24), (25) on the one hand, (26) on the other hand, give two different forms of $S(z)$. Actually, this is an algebraic identity that can be checked directly by expanding (26) and substituting for Q with the help of equation (14) rewritten

$$Q = P - FPF' = (zI - F)P((1/z)I - F') + FP((1/z)I - F') + (zI - F)PF' .$$

(The last equal sign is a mere trivial algebraic identity.)

Anyhow, the product form of S_{yy} is therefore invariant under a change of realization among the equivalent ones, as (24), (25) shows. We may thus use the filter. But we know that

$$\begin{bmatrix} Q_* & S_* \\ S_*' & R_* \end{bmatrix} = \begin{bmatrix} K \\ I \end{bmatrix} R_*(K'I).$$

Let

$$(27) \qquad W_0(z) = I + H(zI - F)^{-1}K,$$

we arrive at

$$S_{yy}(z) = W_0(z)R_*W'(1/z).$$

Therefore, $W(z) = W_0(z)R_*^{1/2}$ will give a factorization of $S_{yy}(z)$ of the form (4). Since F is stable, formula (27) shows that all its poles are inside the unit disk. Moreover, one can check that

(28) $$W_0^{-1}(z) = I - H(zI - F + KH)^{-1}K.$$

Therefore, since $F - KH$ is also stable, W_0^{-1} has its poles in the unit disk, and $W(z)$ is thus a strong factor.

The conclusion is that solving the Riccati equation (22), or equivalently the stationary version of (10), and finding its minimal solution, or maximal for (10), yields a strong factorization of $S_{yy}(z)$. Finding the right solution may be achieved by using the limit value of (10), or of the recurrence deduced from it for (22). Some more calculations, using (13), (27) and (28) allows one to identify the transfer function of the Kalman filter to the Wiener filter, as it had to be.

Let us finally point out that the last formulas allow one to compute the filter without performing a spectral factorization, nor solving a Riccati equation (see Bernhard et al. [3]). The method is based upon the fact that (27), together with the identity we pointed out between the two forms of $S_{yy}(z)$, yield

$$S_{yy}'^{-1}(z) = \bar{\Lambda}_0 + \bar{H}(zI - \bar{F})^{-1}\bar{G} + \bar{G}'((1/z)I - \bar{F}')^{-1}\bar{H}',$$

where

$$\bar{F} = F - KH \quad \text{and} \quad \bar{H} = K.$$

Therefore the algorithm is as follows: Through an FFT compute $S_{yy}(z)$ (one gets it as a set of numerical matrices, corresponding to various values of z on the unit circle), compute $S_{yy}'^{-1}(z)$, perform a new FFT to be back in the time domain, and realize the sequence starting at $t = 1$. (The separation of the sequences for negative and for positive arguments is the equivalent of the spectral factorization, trivially done in the time domain). This yields $F - KH$ and K, thus the filter. Recovering $\hat{y}$ requires that one realize the original sequence $\Lambda(t)$ to get H. Some care must be exerted to get it in the same basis as the two preceding ones.

Kalman filtering is now a classical topic. It has been extended to infinite dimensional systems (Bensoussan [2]). However, we see that for the finite dimensional stationary case, it is now only one piece in a very complete theory that embraces Wiener's filter as well. Time domain and frequency domain are no longer two separate approaches to the filtering problem, one may help the other one, or the two may be used in a single algorithm, as the last method shows. Deciding which technique to use for a particular case depends on other features of the problem: Dimension of y, extra a priori knowledge on the structure of the process, etc.

REFERENCES

1. G. Alengrin and G. Favier, Algorithmes de réalisation stochastique pour l'estimation du gain stationnaire du filtre de Kalman dans le case de systèmes multivariables, to appear.

2. A. Bensoussan, *Filtrage des systèmes linéaires,* Dunod, Paris, 1969.

3. P. Bernhard, A. Benveniste, G. Cohen and J. Chatelon, A new algorithm for Gauss Markov Identification. IFIP Conference on Optimization Techniques, Novosibirsk, USSR, 1974. *Springer Lecture Notes.*

4. P. Faurre, Réalisations Markoviennes de processus stationnaires, *Rapport LABORIA (13):* (1973).

5. P. Faurre, M. Clerget and F. Germain, *Opérateurs rationnels de type positif,* Dunod, Paris, 1979.

6. R. E. Kalman, A new approach to linear filtering and prediction problems. *Journal of Basic Engineering, Trans. ASME Series D, 82D:* (1960) 34-45.

7. R. E. Kalman, *Linear stochastic filtering, theory reappraisal and outlook.* Proceeding of the Symposium on System Theory, Polytechnic Institute of Brooklyn, New York, 1965, 198-205.

8. R. E. Kalman and R. S. Bucy, New results in linear prediction and filtering theory. *Journal of Basic Engineering, Trans. ASME, Series D, 83D:* (1961) 95-100.

9. N. Wiener, *The extrapolation, interpolation and smoothing of stationary time series with engineering applications,* John Wiley and Sons, New York, 1949.

SINGULARITES DE CHAMPS HAMILTONIENS. LIEN AVEC L'EXISTENCE DE SOLUTIONS OPTIMALES EN CALCUL DES VARIATIONS

J. ORTMANS

C.E.R.E.M.A.D.E.
Université de Paris IX Dauphine

RÉSUMÉ

On considère le problème de calcul des variations dans R^n:

$$\text{(P)} \qquad \inf_x \{ \int_0^T f(x,\dot{x})dt;\ x(0) = x_0,\ x(T) = x_1 \}$$

où le critère f est une fonction donnée, régulière mais non convexe par rapport à $\dot{x}$. Les conditions nécessaires d'optimalité donnent lieu à un champ hamiltonien dans l'espace des (x,p), qui présente en certains points des singularités. Dans le cas $n = 1$, I. Ekeland classe les singularités possibles en général et en déduit une condition nécessaire et suffisante sur f pour que le problème (P) admette une solution quelles que soient les conditions aux limites (x_0,x_1,T) ([3]). Dans cet article, on s'intéresse au cas des fonctions critères de "type séparé":

$$f(x,\dot{x}) = g(x) + h(\dot{x}) \qquad (h \text{ non convexe}).$$

Pour $n = 1$, on montre qu'en général le champ hamiltonien ne présente plus certains types de singularités. On étudie alors l'effet d'une perturbation effectuée sur le critère: apparaissent en général de nouveaux types de singularités que l'on classe complètement. Finalement, on détermine les types de singularités qui conduisent à l'absence de solution pour un problème paramétré $(P_\tau)_{\tau\varepsilon[0,1]}$.

Les démonstrations des résultats présentés dans cet article sont développées dans [9]. Ces résultats font partie d'une thèse de troisième cycle dirigée par I. Ekeland. Je lui exprime ma très vive reconnaissance pour sa grande disponibilité et pour les excellents conseils qu'il m'a donnés.

1. Bitrajectoires Associées au Problème (c.f. le Résumé)

Dans ce paragraphe, nous réintroduirons la terminologie et les définitions de [3].

1.1 Les Hypothèses Générales et le Problème

On se fixe la fonction θ: $R^+ \to R$, bornée inférieurement et telle que:

$$\theta(t)/t \mapsto +\infty \quad \text{quand} \quad t \mapsto +\infty .$$

Soit $C^\infty_\theta(R^n \times R^n)$ l'ensemble des fonctions numériques f indéfiniment différentiables sur $R^n \times R^n$ et vérifiant:

$$\text{(E)} \qquad f(x,u) \geq \theta(\|u\|) \quad \forall\ (x,u) \ \varepsilon\ R^n \times R^n.$$

On munit cet espace de la topologie de Whitney $W\infty$ qui en fait un espace de Baire ([5] ou [8]).

Une propriété sera dite vraie pour presque toutes les fonctions $f \ \varepsilon\ C^\infty_\theta(R^n \times R^n)$, ou encore générique, si l'ensemble des fonctions $f \ \varepsilon\ C^\infty_\theta(R^n \times R^n)$ qui la satisfont contient un G_δ dense de $C^\infty_\theta(R^n \times R^n)$.

On considère le problème:

$$\text{(P)} \qquad \inf_x\{ \int_0^T f(x(t),\dot{x}(t))dt;\ x(0) = x_0,\ x(T) = x_1\}$$

où x est une fonction absolument continue de $[0,T]$ dans R^n dont la dérivée $\dot{x}$ appartient à L^1, et où f est une fonction donnée de $C^\infty_\theta(R^n \times R^n)$.

Notons que: -aucune hypothèse de convexité n'est faite sur $f(x,\cdot)$, pour tout $x \in R^n$ fixé; il se peut donc que pour certaines données, le problème (P) n'ait pas de solution;

-le temps ne figure pas explicitement dans le critère $f(x,\dot{x})$.

Soit $p \in R^n$ la variable adjointe à x, $\mathcal{H}$ l'hamiltonien du problème et H l'hamiltonien réduit:

$$\mathcal{H}(x,p;u) = f(x,u) + \langle p,u\rangle \qquad \forall\ (x,p,;u) \in (R^n)^3,$$

$$H(x,p) = \min_{u \in R^n} \mathcal{H}(x,p,;u).$$

Soit alors Ω l'ensemble des (x,p) tels que la fonction $\mathcal{H}(x,p;\cdot)$ atteigne son minimum absolu en un unique point $\hat{u}(x,p)$ et que la forme quadratique f''_{uu} soit non dégénérée en $\hat{u}(x,p)$. Ω est un ouvert de $R^n \times R^n$ sur lequel la fonction H et l'application $\hat{u}$ sont définies et indéfiniment différentiables. Dans Ω, on peut écrire les conditions nécessaires d'optimalité (équations de Pontryagin) sous forme d'un système d'équations du type Hamilton-Jacobi:

$$(S) \qquad \begin{cases} \dot{x} = H'_p(x,p) = \hat{u} \\ \dot{p} = -H'_x(x,p) = -f'_x(x,\hat{u}). \end{cases}$$

Les courbes intégrales définies par ce système constituent un feuilletage de Ω.

On appellera *bisolution* toute application $(x(\cdot),p(\cdot))$ vérifiant le système (S) sur $[0,T]$, et *bitrajectoire* l'image dans $R^n \times R^n$ d'une bisolution.

La condition nécessaire d'optimalité peut donc se formuler géométriquement en disant que toute solution x du problème (P) doit se relever en une bisolution (x,p).

1.2 Partition de $R^n \times R^n$ par les Ensembles Σ_i

On appelle codimension d'une singularité le nombre de paramètres nécessaire à la description complète de la singularité.

En chaque point (x,p) donné, la fonction $\mathcal{V}(x,p;\cdot)$ atteint son minimum en K points distincts u_k, $1 \leq k \leq K \leq \infty$. Chacun d'eux est un point critique de la fonction, qui y présente donc une singularité de codimension c_k. Notons que l'on exclut les singularités qui ne sont pas associées à un minimum, et par conséquent toutes les singularités de codimension impaire. On définit alors la codimension de $\mathcal{V}(x,p;\cdot)$ comme:

$$\operatorname{codim} \mathcal{V}(x,p;\cdot) = K - 1 + \sum_{k=1}^{K} c_k \ .$$

On en déduit une partition de $R^n \times R^n$ en ensembles Σ_i, $1 \leq i \leq \infty$:

$$\Sigma_i = \{(x,p) \in R^n \times R^n / \operatorname{codim} \mathcal{V}(x,p;\cdot) = 1\} \ .$$

D'où le tableau:

$(x,p) \in \Sigma_i$		nombre de points critiques où $\mathcal{V}(x,p;\cdot)$ atteint son minimum	codimensions respectives c_k des points critiques
Σ_0		1	0
Σ_1		2	0 0
$\Sigma_2 = \Sigma_2^b \cup \Sigma_2^c$:	Σ_2^b	1	2
	Σ_2^c	3	0 0 0

Et lorsqu'un point (x,p) appartient à un ensemble Σ_i, $i \geq 3$, la fonction $\mathcal{V}(x,p;\cdot)$ atteint son minimum soit en au moins un point critique dégénéré $(c_k \geq 2)$, soit en au moins quatre points critiques.

1.3 Cas des Fonctions Critère de "Type Séparé"

On considère les fonctions f du type:

$$f(x,u) = g(x) + h(u) \qquad \forall\ (x,u) \in R^n \times R^n .$$

D'après l'estimation (E), la fonction g est bornée inférieurement. Soit Ig sa borne inférieure. On ne modifie pas le problème en translatant les deux fonctions g et h respectivement de (-Ig) et Ig. On peut donc supposer que g appartient à $C^\infty_+(R^n)$, ensemble des fonctions positives indéfiniment différentiables sur R^n, et que h appartient à $C^\infty_\theta(R^n)$, ensemble des fonctions indéfiniment différentiables sur R^n et vérifiant:

$$h(u) \geq \theta(\|u\|) \qquad \forall\ u \in R^n.$$

Chacun de ces espaces est naturellement muni de la topologie de Whitney $W\infty$ qui en fait un espace de Baire, on peut donc encore y parler de généricité.

Dans ce cas l'écriture du hamiltonien est:

$$\mathcal{H}(x,p;u) = g(x) + h(u) + pu \qquad \forall\ (x,p;u) \in (R^n)^3$$

et la fonction $\mathcal{H}(x,p;\cdot)$ atteint son minimum en un (des) point(s) critique(s) u(p) indépendant(s) de x.

Pour n = 1, chaque ensemble Σ_i, $1 \leq i \leq \infty$, est donc constitué par une union, éventuellement vide, de droites dont les équations sont du type: p = constante.

Une bitrajectoire est une courbe de $R \times R$, et son intersection avec Σ_i est un phénomène stable que nous allons analyser. Soit:

(x,p) un point de Σ_i ;

u_1 et u_2 les points où $\mathcal{H}(x,p;\cdot)$ atteint son minimum;

$\vec{v}_i = (u_i,-g'_x(x))$, $i = 1,2$, les deux vecteurs vitesses associés;

$\vec{n} = (0,1)$ le vecteur normal à Σ_1 .

Il apparaît que l'ensemble Σ_1 sépare un voisinage $\mathcal{O}$ de (x,p) en deux régions Ω_1 et Ω_2 telles que si $(y,q) \in \Omega_i$, $i = 1,2$, $\mathcal{H}(y,q;\cdot)$ atteint son minimum absolu en $u_i(q)$.

Au point (x,p), on vérifie aisément la conversation de la vitesse normale:

$$\vec{n} \cdot \vec{v}_1 = \vec{n} \cdot \vec{v}_2 .$$

Si bien que deux cas seulement sont possibles à la traversée de Σ_1:

-arrivée transversale - départ transversal: $\vec{n} \cdot \vec{v} \neq 0$,

-arrivée tangentielle - départ tangentiel: $\vec{n} \cdot \vec{v} = 0$.

Dans le premier cas $(\vec{n} \cdot \vec{v} \neq 0)$, au passage de Σ_1, la vitesse saute de $\vec{v}_1$ à $\vec{v}_2$, et la bitrajectoire présente une brisure au point (x,p). On remarque que Σ_2^c joue un rôle analogue à celui de Σ_1.

Une *bitrajectoire* sera dite *générique* si elle est contenue dans $\Sigma_0 \cup \Sigma_1$, ses intersections avec Σ_1 vérifiant la condition $\vec{n} \cdot \vec{n} \neq 0$. Sinon, elle sera dite *singulière*. Une bitrajectoire sera donc singulière:

-soit parce qu'elle ne vérifie pas la condition $\vec{n} \cdot \vec{v} \neq 0$ aux points de contact avec Σ_1;

-soit parce qu'elle recontre des ensembles Σ_i, $i \geq 2$.

Nous allons étudier tout d'abord deux types de singularités:

les singularités de type I qui sont les points de Σ_1 où $g'(x) = 0$.

et les singularités de type II qui sont les points de Σ_2.

2. Singularités de Types I et II

Le critère, c'est-à-dire le couple $(g,h) \in C_+^\infty(R) \times C_\theta^\infty(R)$, est donné, on lui associe le hamiltonien réduit H.

Commençons par faire les trois remarques suivantes:

REMARQUE 2.1. Soit (x,p) un point appartenant à Σ_1, on a:

$$d(\vec{n}\cdot\vec{v}_1)r/dt^r = d(\vec{n}\cdot\vec{v}_2)r/dt^r = -g^{(r+1)}(x) \qquad \forall\ r \geq 0 .$$

Et on obtient des égalités semblables pour décrire tout contact de la bitrajectoire avec un ensemble Σ_i, $i \geq 2$.

REMARQUE 2.2. Génériquement, $q \in C_+^\infty(R)$ est une fonction de Morse ([1]).

REMARQUE 2.3. On *pressent* que, génériquement, $u_i(p) \in R^*$, $i = 1,2$. En effet, dire que l'un des $u_i(p)$ est nul se traduit par la situation géométrique suivante:

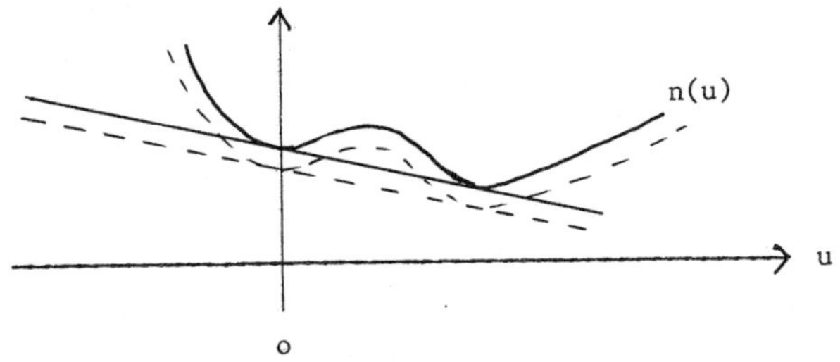

Et il suffit de translater un peu la fonction h pour avoir $u_i(p) \neq 0$, $i = 1,2$.

Par suite, nous dirons que le champ hamiltonien, noté également H, présente au point (x,p) *une singularité générique de type* I s'il existe exactement deux réels u_1 et u_2 tels que:

$$
\text{I} \quad \begin{cases} h(u_1) + pu_1 = h(u_2) + pu_2 \\ \left.\begin{array}{l} h'(u_i) + p = 0 \\ h''(u_i) > 0, \quad u_i \neq 0 \end{array}\right| \quad i = 1,2 \\ g'(x) = 0, \quad g'(x) \neq 0 \quad (\dagger) \end{cases}
$$

Cette singularité est de codimension 0.

Définissons maintenant les singularités de type I et de codimension 1. Nous les appellerons singularités de transition en raison du caractère en général instable de leur apparition que nous étudierons au paragraphe 3. Il y en a seulement deux types notés TI et T_0I.

On dira que le champ hamiltonien H présente au point (x,p) une *singularité de type* TI *(respectivement* T_0I*)* s'il existe exactement deux réels u_1 et u_2 tels que:

$$
\text{TI} \quad \begin{cases} h(u_1) + pu_1 = h(u_2) + pu_2 \\ \left.\begin{array}{l} h'(u_i) + p = 0 \\ h''(u_i) > 0, \quad u_i \neq 0 \end{array}\right| \quad i = 1,2 \\ g'(x) = g''(x) = 0, \quad g^{(3)}(x) = 0 \quad (\dagger) \end{cases}
$$

(†) Cf. Remarque 2.1.

$$T_0 I \quad \begin{cases} h(u\ ) + pu\ = h(u\ ) + pu \\ h'(u\ ) + p = 0, \quad h''(u\ ) > 0 \qquad i = 1,2 \\ u_1 = 0, \quad u_2 \neq 0 \ ^{(\dagger)} \\ g'(x) = 0, \ g''(x) \neq 0 \ _{(\dagger\dagger)} \end{cases}$$

Nous avons vu que les singularités de type II sont de deux types. Compte tenu des remarques 2.1, 2.2 et 2.3, on dira que le champ hamiltonien H présente au point (x,p) une *singularité générique de type* IIb *(respectivement* IIc*)* s'il existe exactement un réel $\bar{u}$ (respectivement trois réels u_1, u_2, u_3) tel que:

$$\text{IIb} \quad \begin{cases} h(\bar{u}) + p\bar{u} = \min\limits_{v} (h(v) + pv) \\ h'(\bar{u}) + p = h''(\bar{u}) = h^{(3)}(\bar{u}) = 0 \\ h^{(4)}(\bar{u}) > 0, \quad \bar{u} \neq 0 \\ q \text{ est une fonction de Morse} \end{cases}$$

$$\text{IIc} \quad \begin{cases} h(u_1) + pu_1 = h(u_2) + pu_2 = h(u_3) + pu_3 \\ \left. \begin{array}{l} h'(u_i) + p = 0 \\ h'(u_i) > 0, \quad u_i \neq 0 \end{array} \right| \quad i = 1,2,3 \\ q \text{ est une fonction de Morse} \end{cases}$$

Avec les singularités de types TI et $T_0 I$, ce sont les seules singularités de codimension 1.

(†) Cf. Remarque 2.1.

(††) Ou bien $u_1 \neq 0$, $u_2 = 0$.

PROPOSITION 2.1. Pour presque tous les couples $(g,h) \in C_+^\infty(R) \times C_\theta^\infty(R)$, le champ hamiltonien associé ne présente que des singularités génériques de type I.

REMARQUE 2.4. Lorsque le critère est de type "non-séparé," le champ hamiltonien peut présenter en général des singularités génériques de types I et II ([3]).

COROLLAIRE 2.1. Pour presque toutes les fonctions $h \in C_\theta^\infty(R)$, les ensembles Σ_i, $i \geq 2$ sont vides.

D'après la description des ensembles Σ_i pour $n = 1$ (§1.2), si génériquement Σ_2 est vide, il en sera de même pour Σ_i, $i \geq 3$. Ainsi pour démontrer la Proposition 2.1, (et le corollaire), il suffira donc de montrer que, génériquement, un champ hamiltonien ne présente aucun des quatre types de singularités: TI, T_0I, IIb, IIc. Cette démonstration repose sur des arguments de transversalité.

Nous allons maintenant étudier l'effet d'une perturbation sur la fonction critère.

3. Paramètrage de la Fonction Critère

3.1 Naissance et Mort des Singularités de Codimension 1

Perturbons la fonction critère en introduisant un paramètre reél τ. Le couple (g,h) qui était fixé s'écrit maintenant $({}_\tau g, {}_\tau h)$ et évolue continûment en fonction des valeurs de τ. On peut faire apparaître pour une certaine valeur du paramètre, un des quatre types de singularités de codimension 1.

La disparition et l'apparation d'une singularité de type T_0I ont déjà été évoquées dans la Remarque 2.3.

Voici par exemple une autre évolution possible: la fonction g est strictement convexe et on considère un paramétrage qui n'affecte que la fonction h (Figure 3.1).

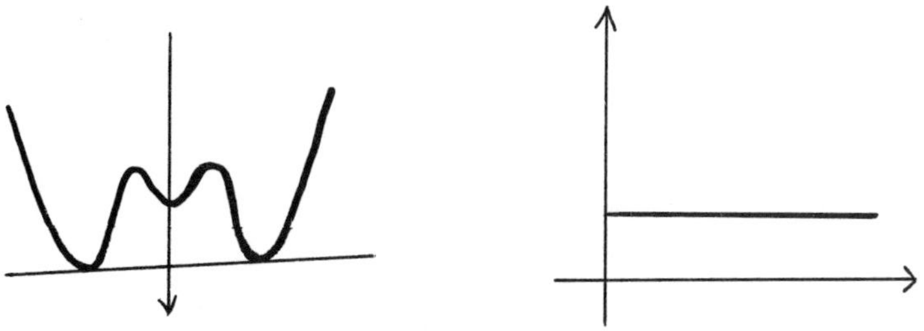

Une singularité générique de Type I

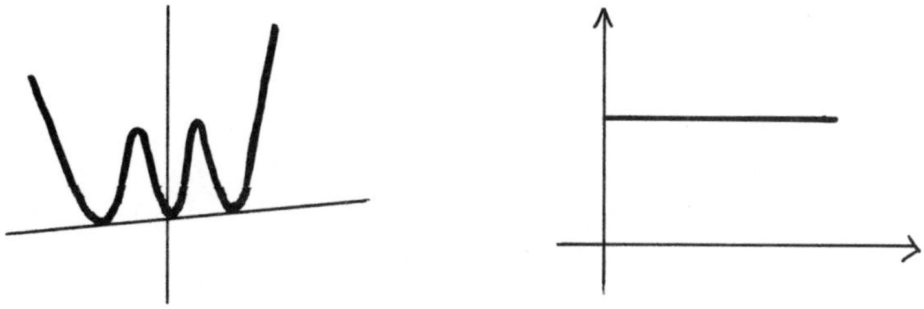

Une Singularité de Type IIC

"Eclatement de Σ_2^c"

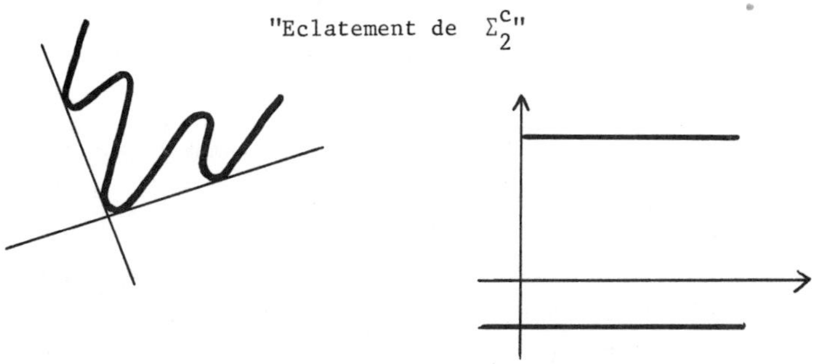

Deux singularités génériques de Type I

"Eclatement de Σ_2^c"

FIGURE 3.1

Nous allons déterminer si, lors de ces évaluations, on peut voir apparaître de nouveaux types de singularités.

3.2 Paramétrage du Couple-Critère

Considérons les chemins, notés (g,h), indéfiniment différentiables dans $C_+^\infty(R) \times C_\theta^\infty(R)$

$$(g,h): [0,1] \to C_+^\infty(R) \times C_\theta^\infty(R)$$

$$\tau \to ({}_\tau g, {}_\tau h): (x,u) \to {}_\tau g(x) + {}_\tau h(u) .$$

Soit: ${}_\tau g(x) = g(\tau,x)$

${}_\tau h(u) = h(\tau,u)$.

Nous considérons maintenant g et h comme des fonctions de deux variables:

$$g \in C_+^\infty([0,1] \times R)$$

$$h \in C_\theta^\infty([0,1] \times R).$$

Les deux espaces sont munis de la topologie de Whitney $W\infty$. Par suite le chemin (g,h) appartient à l'espace $C_+^\infty([0,1] \times R) \times C_\theta^\infty([0,1] \times R)$ muni de la topologie produit induite.

Soit H_τ le hamiltonien réduit associé au couple critère $({}_\tau g, {}_\tau h)$. On associe au chemin (g,h) une "*variation de champs hamiltoniens*" notée H et telle que:

$$H(\tau x,p) = H_\tau(x,p) \qquad \forall\ (\tau,x,p) \in [0,1] \times R^2.$$

Nous dirons qu'une variation de champs hamiltoniens H présente une singularité de type "X" s'il existe un triplet $(\tau,x,p) \in [0,1] \times R^2$ tel que le champ hamiltonien H_τ présente au point (x,p) une singularité de ce type.

THÉORÈME 3.1. Pour presque tous les chemins $(g,h) \in C_+^\infty([0,1] \times R) \times C_\theta^\infty([0,1] \times R)$, la variation de champs hamiltoniens H associée ne présente que des singularités de types I, TI, T_0I, IIb, IIc. De plus, la variation H présente les singularités de codimension 1 pour un nombre fini de valeurs du paramètre τ; et outre les singularités génériques de type I, pour chaque $\tau \in [0,1]$, le champ H_τ presénte des singularités d'un seul des quatre autres types au plus.

Pour démontrer ce théorème, on commence par répartir tous les types de singularités (dans la description desquels intervient maintenant le paramètre τ) en trois ensembles disjoints:

-l'ensemble des singularitiés génériques de type I:

-l'ensemble des singularités de type TI, T_0I, IIb, IIc;

-un ensemble comprenant douze familles S_j $(1 \leq j \leq 12)$ de types de singularités, naturellement déduites des ensembles précédents.

Ensuite, l'utilisation des théorèmes de transversalité ([5],[6] ou [7]) pour chaque type de singularités ou pour chaque famille S_j $(1 \leq j \leq 12)$ nous conduit à analyser essentiellement trois situations qui finalement impliquent les résultats du théorème.

REMARQUE 3.1. On peut étudier de façon analogue le paramétrage d'une fonction critère non de type séparé.

Classons maintenant les singularités génériquement présentées par une variation de champs hamiltoniens.

4. Allure de Champ Hamiltonien au Voisinage des Singularités de Codimension ≤ 1

Le paramètre τ a maintenant une valeur déterminée, nous ne le ferons donc pas figurer dans les écritures.

Pour tous les résultats qui vont suivre, on suppose que:

- le type de singularités étudié est présenté au point (x,p);

- la fonction $\mathcal{K}(x,p;\cdot)$ atteint son minimum en un (des) point(s) critique(s) $u_i(p)$ noté(s) u_i; si (x,p) est un point de Σ_2^c, on suppose: $u_1 < u_3 < u_2$;

- la valeur de la première dérivée non nulle de la fonction g en x est notée b.

Indiquons le principe des démonstrations en considérant, par exemple, un point $(x,p) \in \Sigma_1$. Rappelons que Σ_1 est une droite d'équation p = constante qui sépare localement l'espace en deux régions Ω_1 et Ω_2, (§1.3), avec $\Omega_1 \cap \Omega_2 = \Sigma_1$.

Et on a:

$$\forall\ (y,q) \in \Omega_i,\quad H(y,q) = \mathcal{K}(y,q;u_i(q));\quad i = 1,2\ .$$

Introduisons la fonction ν_i suivante:

$$\nu_i(q) = h(u_i(q)) + q\ u_i(q)\quad \forall\ (y,q) \in \Omega_i;\quad i = 1,2.$$

On a naturellement:

$$\nu_1(p) = \nu_2(p).$$

Le hamiltonien réduit s'écrit alors:

$$H(y,q) = g(y) + \nu_i(q)\quad \forall\ (y,q) \in \Omega_i;\quad i = 1,2\ .$$

L'équation H = constante caractérise complètement les bitrajectoires. Les développements respectifs des fonctions g et ν dans un voisinage assez petit du point (x,p) considéré et des changements de coordonnées corrects permettent de déterminer les équations des bitrajectoires dans un modèle local.

Nous appelons modèle local la donnée d'un système de coordonnées locales, au voisinage du point $(x,p) \in \Sigma_i$, ayant pour axes Σ_i et $\vec{n}$ $(i = 1,2,\ldots)$. Pour chaque modèle, les systèmes de changement de variables définissent un difféomorphisme local ϕ d'un voisinage $\mathcal{O}$ de (x,p) sur R^2 de classe C^∞ et tel que:

$$H(y,q) = L \circ \phi(y,q) = L(\xi,\pi) \quad \forall\ (y,q) \in \mathcal{O}.$$

Les singularités génériques de type I (codimension 0) ont déjà été classées dans [3]:

THÉORÈME D'EKELAND: Il y a trois classes de singularités génériques de type I. Pour chacune d'elles on dispose d'un modèle local $(L,0,0)$:

$$\text{I a:} \quad 0 \notin [u_1,u_2] \quad ; \; L(\xi,\pi) = \pi + \xi^2 + c$$

$$\text{I b:} \quad 0 \in]u_1,u_2[, \; b < 0; \; L(\xi,\pi) = -|\pi| - \xi^2 + c$$

$$\text{I c:} \quad 0 \in]u_1,u_2[, \; b > 0; \; L(\xi,\pi) = -|\pi| + \xi^2 + c$$

où c est une constante.

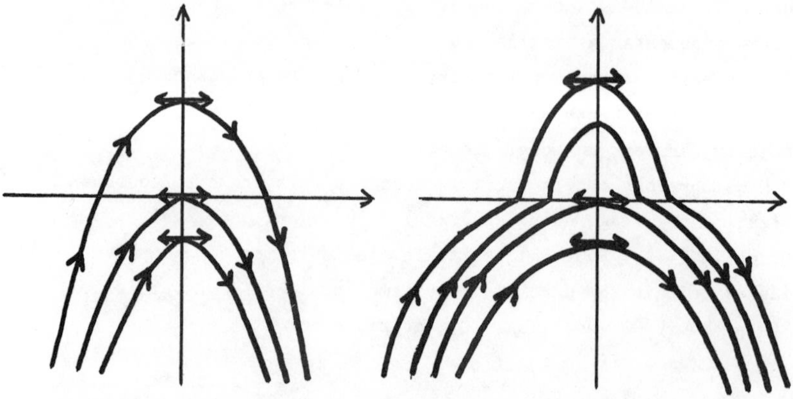

Modèle local ($b > 0$, $0 < u_1 < u_2$) *Réalité* ($|u_1| < |y_2|$)

<u>Singularité de type Ia</u>

Notons que lorsque l'on connait l'allure des bitrajectoires dans le modèle local, on en déduit aisément leur allure dans la réalité. Pour toute la suite, seules les configurations géométriques du champ *dans les modèles locaux* seront donc représentées.

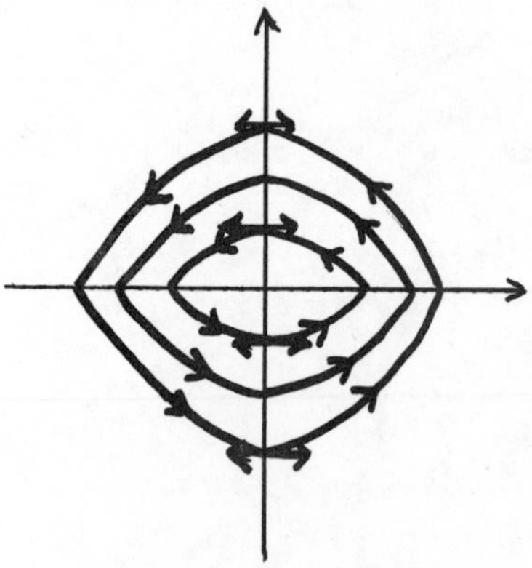

Singularité de type Ib

$b < 0$

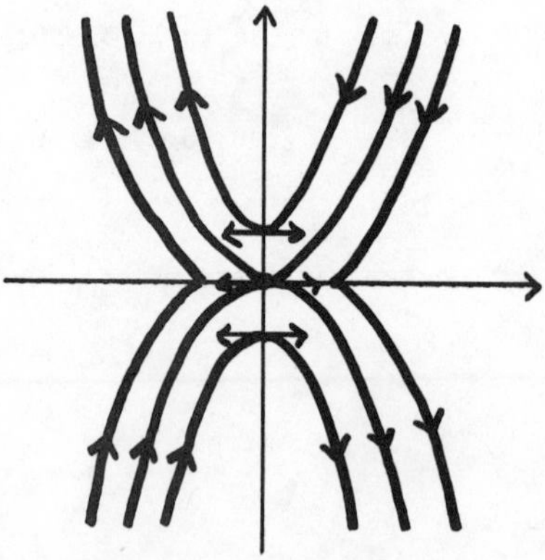

Singularité de type Ic

$b > 0$

FIGURE 4.1

Classons maintenant les types de singularités de codimension 1.

PROPOSITION 4.1. Il y a deux classes de singularités de type TI. Pour chacune d'elles on dispose d'un modèle local (L,0,0):

$$\text{TI aa:} \quad 0 \notin [u_1,u_2]; \quad L(\xi,\pi) = \pi + \xi^2 + c$$

$$\text{TI bc:} \quad 0 \in]u_1,u_2[; \quad L(\xi,\pi) = -|\pi| + \xi^3 + c$$

où c est une constante.

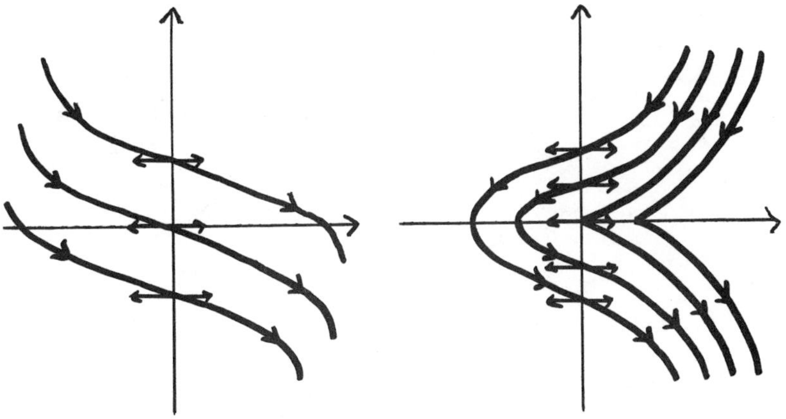

Singularité de type TIaa
$b > 0,\ 0 < u_1 < u_2$

Singularité de type TIbc
$b > 0$

FIGURE 4.2

PROPOSITION 4.2. Il y a deux classes de singularités de type T_0I. Pour chacune d'elles on dispose d'un modèle local (L,0,0):

$$T_0I \text{ ab:} \quad b < 0; \ L(\xi,\pi) = \begin{cases} -\pi^2 - \xi^2 + c \text{ pour } \pi > 0 \\ \pi - \xi^2 + c \text{ pour } \pi < 0 \end{cases}$$

$$T_0Iac: \quad b > 0; \; L(\xi,\pi) = \begin{cases} -\pi^2 + \xi^2 + c & \text{pour } \pi > 0 \\ \pi + \xi^2 + c & \text{pour } \pi < 0 \end{cases}$$

où c est une constante.

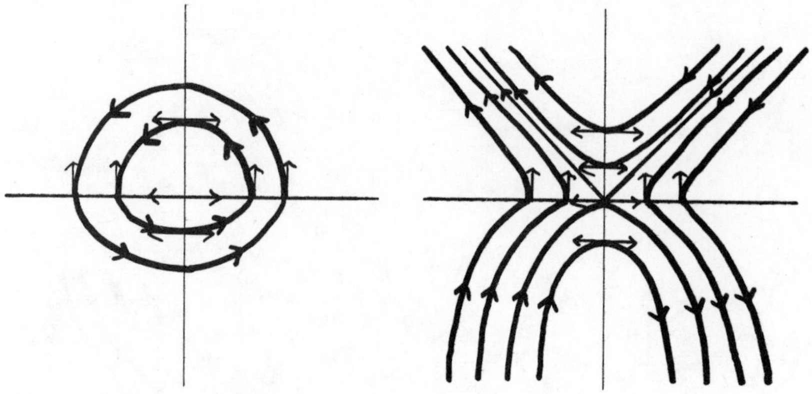

Singularité de type T_0Iab *Singularité de type T_0Iac*

FIGURE 4.3

PROPOSITION 4.3. Il y a deux classes de singularités génériques de type IIb. Pour chacune d'elles on dispose d'un modèle local $(L,0,0)$:

$$IIb_1: \quad g'(x) \neq 0; \quad L(\xi,\pi) = \pi + \xi + c$$

$$IIb_2: \quad g'(x) = 0; \quad L(\xi,\pi) = \pi + \xi^2 + c$$

où c est une constante.

PROPOSITION 4.4. Il y a cinq classes de singularités génériques de type IIc. Pour chacune d'elles on dispose d'un modèle local $(L,0,0)$:

$$IIc_{11}: \; 0 \notin [u_1,u_2], \quad g'(x) \neq 0; \quad L(\xi,\pi) = \pi + \xi + c$$

$$\mathrm{IIc}_{21}: \quad 0 \notin [u_1,u_2], \begin{cases} g'(x) = 0; & L(\xi,\pi) = \pi + \xi^2 + c \\ g''(x) \neq 0 \end{cases}$$

$$\mathrm{IIc}_{12}: \quad 0 \in]u_1,u_2[, \quad g'(x) \neq 0; \quad L(\xi,\pi) = -|\pi| + \xi + c$$

$$\mathrm{IIc}_{22}: \quad 0 \in]u_1,u_2[, \begin{cases} g'(x) = 0; & L(\xi,\pi) = -|\pi| + \xi^2 + c \\ g''(x) > 0 \end{cases}$$

$$\mathrm{IIc}_{23}: \quad 0 \in]u_1,u_2[, \begin{cases} g'(x) = 0; & L(\xi,\pi) = -|\pi| - \xi^2 + c \\ g''(x) < 0 \end{cases}$$

où c est une constante.

Dans le modèle local respectivement associé à chaque type de singularités, un champ hamiltonien a la même configuration géométrique au voisinage d'une singularité de type IIb_2 (respectivement IIc_{21}, IIc_{22}, IIc_{23}) et au voisinage d'une singularité de type Ia (respectivement Ia, Ic, Ib).

Dans le modèle local associé, au voisinage d'une singularité de type IIb_1 ou IIc_{11}, les bitrajectoires sont des droites parallèles à la seconde bissectrice. Dans chaque cas il est naturel que la bitrajectoire passant par 0 ne présente aucune particularité par rapport à celles passant par les points voisins de l'axe des ξ, car on voit au cours de la démonstration du théorème 3.1 que ce type de singularité n'est pas présenté en des points (x,p) isolés.

Dans le cas d'une singularité de type IIc_{12}, les bitrajectoires vérifient simplement la "loi de la réfraction" ([13]) sur Σ_2^c:

$$\vec{v}_2 = \vec{v}_1 + M(\vec{n})$$

où M est la matrice symplectique:

$$M = \begin{pmatrix} 0 & 1 \\ -1 & 0 \end{pmatrix} .$$

Notons enfin que dans la réalité les bitrajectoires présentent une brisure sur Σ_2^c et non Σ_2^b.

Voici un tableau indiquant les possibilités d'existence des singularités de transition et de type II lorsque l'on parcourt des chemins génériques, en fonction des classes de singularités génériques de type I présentées par les variations de champs hamiltoniens respectivement associées.

Type de singularité présenté par la variation de champs hamiltoniens: $\overset{\rightarrow}{\downarrow}$	Pas de singularité	Ia	Ib	Ic
Pas de singularité	$IIc_{11}IIc_{12}$ IIb_1	IIb_2 TIaa	IIb_2 TI	IIb_2 bc(†)
Ia	IIb_2 TIaa	IIc_{21} TIaa	$IIc_{21}IIc_{23}$ T_0Iab	$IIc_{21}IIc_{22}$ T_0Iac
Ib	IIb_2 TIbc(†)	$IIc_{21}IIc_{23}$ T_0Iab	$IIc_{21}IIc_{23}$ TI	bc(†)
Ic	IIb_2	$IIc_{21}IIc_{23}$ T_0I_{ac}		$IIc_{21}IIc_{22}$

(†) La variation de champs hamiltoniens présente, pour certaines valeurs du paramètre $\tau \in [0,1]$, à la fois une singularité de type Ib et une singularité de type Ic.

Rappelons qu'étant donnés un chemin générique et une valeur τ du paramètre, le champ H_τ associé peut présenter (cf. théorème 3.1):

- des singularités des trois types Ia, Ib, Ic;

- des singularités d'un seul des quatre types TI, T_0I, IIb, IIc au plus.

L'examen des différents modèles locaux va nous permettre de déterminer les types de singularités dont l'existence conduit à l'absence de solutions pour un "problème paramétré" $(P_\tau)_{\tau\varepsilon[0,1]}$.

5. Existence des Solutions

Récrivons le problème (P):

$$(P) \qquad \inf_x \left\{ \int_0^T \{g(x) + h(\dot{x})\}dt;\ \dot{x} \in L^1;\ x(0) = x_0,\ x(T) = x_1 \right\} .$$

Si la fonction h est strictement convexe, tous les ensembles Σ_i, $i \geq 1$, sont vides, et le problème (P) admet une solution pour tout triplet (x_0,x_1,T). Pour déterminer dans quelles conditions le problème n'a pas de solution, on va donc comparer, à l'aide des modèles locaux, les bisolutions des systèmes hamiltoniens associés respectivement au problème (P) et au problème relaxé (P^{**}) qui lui correspond:

$$(P^{**}) \qquad \min_x \left\{ \int_0^T \{g(x) + h^{**}(\dot{x})\}dt;\ \dot{x} \in L^1;\ x(0) = x_0,\ x(T) = x_1 \right\}$$

où h^{**} est la régularisée convexe semi-continue inférieurement de la fonction h. La fonction h^{**} est donc de classe C^∞.

Soit $C(p)$ et $C^{**}(p)$ les ensembles, non vides, tels que:

$$C(p) = \{u \in R / h(u) + pu = \min_v (h(v) + pv)\}$$

$$C^{**}(p) = \{u \in R / h^{**}(u) + pu = \min_v (h^{**}(v) + pv)\} .$$

Il est clair que $C^{**}(p)$ est l'enveloppe convexe de $C(p)$. Soit encore, si (x,p) est un point de l'ensemble Σ_k où la fonction $\mathcal{H}(x,p;\cdot)$ atteint son minimum en j points distincts $u_1,\dots,u_j$, $j \in [1,k+1]$:

$$C^{**}(p) = co\{u_1,\dots,u_j\} .$$

Les conditions nécessaires d'optimalité pour le problème (P^{**}) s'écrivent alors ([2]):

$$(S^{**}) \begin{cases} \dot{x} \in C^{**}(p) \\ \\ \dot{p} = -g'(x) \end{cases} .$$

Et toute bisolution du système (S) est une bisolution du système (S^{**}), comme l'indique le théorème de relaxation ([4]) dont un corollaire est:

COROLLAIRE 5.1. L'ensemble (non vide) des solutions du problème relaxé contient l'ensemble (éventuellement vide) des solutions du problème originel. De plus les deux problèmes ont même valeur:

$$\inf_x \{ \int_0^T \{g(x) + h(\dot{x})\}dt\} = \min_x \{ \int_0^T \{g(x) + h(\dot{x})\}dt\}$$

pour $x(0) = x_0$ et $x(T) = x_1$.

Du théorème d'existence d'Ekeland ([4],§5), on déduit:

COROLLAIRE 5.2. Pour presque tous les couples critères $(g,h) \in C_+^\infty(R) \times C_\theta^\infty(R)$, une condition nécessaire et suffisante pour que le problème (P) admette au moins une solution quelles que soient les conditions aux limites (x_0,x_1,T) est que le champ hamiltonien associé ne présente pas de singularité générique de type Ic.

Considérons maintenant le cas d'un *problème paramétré*, c'est à dire d'un ensemble de problèmes $(P_\tau)_{\tau \in [0,1]}$ associé à un chemin $(g,h) \in C_+^\infty([0,1] \times R) \times C_\theta^\infty([0,1] \times R)$ (§3.2):

$$(P_\tau)_{\tau \in [0,1]}: \quad \inf_x \{ \int_0^T \{g(\tau,x) + h(\tau,\dot{x})\, dt;\ \dot{x} \in L^1;\ x(0) = x_0,\ x(t) = x_1\} .$$

Nous dirons qu'un problème paramétré admet une solution pour tout triplet (x_0,x_1,T) si, pour chaque $\tau \in [0,1]$, le problème (P_τ) admet une solution pour tout triplet (x_0,x_1,T).

THÉORÈME 5.1. Pour presque tous les chemins $(g,h) \in C_+^\infty([0,1] \times R) \times C_\theta^\infty([0,1] \times R)$ une condition nécessaire et suffisante pour que le problème paramétré $(P_\tau)_{\tau \in [0,1]}$ associé admette une solution quelles que soient les conditions aux limites (x_0,x_1,T) est que la variation de champs hamiltoniens correspondante ne présente pas de singularité générique de type Ic.

Avant de considérer la démonstration de ce théorème, énonçons quelques résultats préliminaires.

DÉFINITION 5.1. Nous noterons:

- G l'ensemble des fonctions $g \in C_+^\infty(R)$ que présentent au moins au minimum en un point critique non dégénéré $x \in R$.

- M l'ensemble des fonctions $h \in C_\theta^\infty(R)$ pour lesquelles il existe un réel p tel que la fonction $u \to h(u) + pu$ associée atteigne son minimum en deux points critiques non dégénérés, non nuls, et de signes opposés.

- E l'ensemble des couples critères $(g,h) \in C_+^\infty(R) \times C_\theta^\infty(R)$ tels que le champ hamiltonien associé ne présente que des singularités de codimension ≤ 1.

- E_s le sous ensemble de E constitué des couples critères tels que le problème (P) associé admette au moins une solution quelles que soient les conditions aux limites (x_0,x_1T).

On montre que l'ensemble G est ouvert mais n'est pas dense dans $C_+^\infty(R)$, pour la topologie de Whitney $W\infty$. Par suite, pour presque tous les couples $(g,h) \in C_+^\infty(R) \times C_\theta^\infty(R)$:

- la fonction g étant de Morse: soit elle ne présente pas de minimum, soit elle appartient à l'ensemble G;

- dire que le champ hamiltonien présente une singularité de type Ic, c'est dire que le couple (g,h) appartient à l'ensemble $G \times M$. En outre, d'après les théorèmes de transversalité, les conditions que doit vérifier une fonction h pour appartenir à l'ensemble M sont stables. L'ensemble M est donc ouvert dans $C_\theta^\infty(R)$.

Ainsi, les conditions qui définissent une singularité générique de type Ic sont stables, et l'ensemble (non vide) des couples critères tels que le champ hamiltonien associé présente une telle singularité est ouvert dans $C_+^\infty(R) \times C_\theta^\infty(R)$. On en déduit un corollaire du théorème 5.1:

COROLLAIRE 5.3. Pour presque tous les chemins $(g,h) \in C_+^\infty([0,1] \times R) \times C_\theta^\infty([0,1] \times R)$, si, sauf pour un nombre fini de valeurs du paramètre τ, on sait que les problèmes (P_τ) associés admettent chacun au moins une solution pour tout triplet (x_0,x_1,T), on en déduit alors que le problème paramétré correspondant admet une solution quelles que soient les conditions aux limites (x_0,x_1,T).

De plus, l'ensemble G étant ouvert et une singularité de type IIc_{22} ne pouvant être présentée que pour des valeurs isolées du paramètre τ (Figure 3.1 et théorème 3.1), on a le corollaire suivant:

COROLLAIRE 5.4. Pour presque tous les chemins $(g,h) \in C_+^\infty([0,1] \times R) \times C_\theta^\infty([0,1] \times R)$, si la variation de champs hamiltoniens associée présente une singularité de type IIc_{22}, alors elle présente aussi des singularités génériques de type Ic.

Nous dirons qu'un couple de fonctions est générique si chacune des fonctions qui le composent ne vérifie que des propriétés génériques.

Ainsi tout chemin générique $(g,h) \in C_+^\infty([0,1] \times R) \times C_\theta^\infty([0,1] \times R)$ est constitué de couples $({}_\tau g, {}_\tau h)$, $\tau \in [0,1]$, appartenant à l'ensemble E.

Principe de démonstration du théorème 5.1.

La condition nécessaire figure déjà dans le corollaire 5.2. Reste à démontrer la condition suffisante. Etant donné un chemin générique, considérons donc un couple critère $(g,h) \in E$ lui appartenant et tel que le champ hamiltonien H associé ne présente pas de singularité de type Ic (d'après le corollaire 5.4, il ne présente pas non plus de singularité de type IIc_{22}). Il suffit maintenant de démontrer que le problème (P) correspondant admet au moins une solution pour tout triplet (x_0,x_1,T).

De même que $\vec{v} = (\dot{x},\dot{p})$ $(\vec{v}_i;\ i = 1,2)$ représente le vecteur vitesse du champ hamiltonien H au point (x,p); notons $\vec{v}_r = (\dot{x},\dot{p})$ *un* vecteur vitesse, au point (x,p) du champ hamiltonien H^{**} associé au problème relaxé.

1ère partie Si $(\bar{x},\bar{p})$ est un point de Σ_0, l'ensemble $C^{**}(\bar{p})$

est réduit à une unique valeur $\dot{x}$, et il passe donc par le point $(\bar{x},\bar{p})$ une unique bisolution locale du système $S(^{**})$ qui est aussi une bisolution locale du système (S).

L'unicité s'entend modulo le déplacement de l'origine des temps.

Si le champ hamiltonien ne présente que des singularités de types Ia, TIaa, IIc_{11}, IIc_{12}, ou IIc_{21}, en un point $(\bar{x},\bar{p})$ non dans Σ_0, tout vecteur vitesse $\vec{v}_r$ est non nul. L'examen des différents modèles locaux montre que toute bisolution de (S^{**}) traverse $(\bar{x},\bar{p})$, elle "repasse" immédiatement dans Σ_0. Par suite, pour toute courbe $x(\cdot)$ associée à une bisolution de (S^{**}), on a:

$$\int_0^T \{g(x(t)) + h(\dot{x}(t))\}dt = \int_0^T \{g(x(t)) + h^{**}(\dot{x}(t))\}dt .$$

Or, d'après le théorème de relaxation, les deux problèmes (P) et (P^{**}) ont même valeur, toute courbe $x(\cdot)$ solution de l'un est donc aussi solution de l'autre.

Si $(\bar{x},\bar{p})$ est une singularité de type T_0I $(u_1 = 0)$, l'unique vecteur vitesse $\vec{v}_r$ nul est le vecteur $\vec{v}_1$. Et l'analyse précédente implique la même conclusion.

2eme partie Cas des singularités de type TIbc, Ib, IIc_{23}.

REMARQUE 5.1. L'examen du modèle local révèle qu'aucune bisolution de (S) ne passe par un point $(\bar{x},\bar{p})$ où le champ hamiltonien présente une singularité de type Ib.

REMARQUE 5.2. Un champ hamiltonien que présente une singularité de type TIbc ou Ib ne présente pas de singularité de type IIc_{23}.

Supposons maintenant que le champ hamiltonien présente une singularité de type TIbc au point $(\bar{x},\bar{p})$. On montre que le modèle local (figure 4.2) peut être considéré comme un modèle global. L'examen de ce modèle indique qu'une unique bisolution B_0 de (S) traverse le point $(\bar{x},\bar{p})$, car le vecteur $\vec{v}_i$ $(i = 1,2)$ est non nul. Mais il existe en ce point un vector $\vec{v}_r = \vec{0}$, des bisolutions de (S^{**}) peuvent donc y séjourner.

Choisissons un triplet (x_0,x_1,T) tel qu'il existe une bisolution B_r de (S^{**}) qui *séjourne* au point $(\bar{x},\bar{p})$ et à laquelle est associée une fonction $x_r(\cdot)$ qui vérifie ces conditions aux limites du problème (P). L'examen du modèle global montre qu'il existe une bisolution B_α (elle couple la droite D d'équation $p = \bar{p}$ au point $(\bar{x} + \alpha,\bar{p})$) solution de (S), ne passant pas par une singularité de type Ib d'après la remarque 5.1, et à laquelle correspond une fonction $x_\alpha(\cdot)$ safisfaisant également les conditions imposées. Soit:

$$I_r = \int_0^T \{g(x_r(t)) + h^{**}(\dot{x}r(t))\}dt$$

$$I = \int_0^T \{g(x_\alpha(t)) + h^{**}(\dot{x}_\alpha(t))\}dt .$$

Avec une méthode de calcul différente pour chaque expression, on exprime (sous des forme comparables) I_r en fonction de la valeur constante $H(\bar{x},\bar{p})$ du hamiltonien réduit sur la bisolution B_r et I_α en fonction des valeurs $H(\bar{x} + \Gamma,\bar{p})$ correspondant respectivement aux bisolutions B_Γ $(D \cap B_\Gamma = (\bar{x} + \Gamma,\bar{p});\ \Gamma \in [0,\alpha])$ de (S) (donc de (S^{**})).

Or, les formules de changement de variables écrites au cours des démonstrations du paragraphe 4 indiquent que $g(\bar{x})$ est un maximum strict absolu de $g(\bar{x} + \Gamma)$ pour $\Gamma \in]0,\alpha]$. On en déduit que $I\alpha$ est strictement inférieur à I_r.

Par suite, la fonction $x_r(\cdot)$ n'est pas une solution du problème relaxé (P^{**}), les deux problèmes (P) et (P^{**}) ont donc les mêmes solutions quelles que soient les conditions aux limites (x_0,x_1,T).

Quant au cas des singularités de types Ib et IIc_{23}, ils sont semblables car ces singularités ont même modèle local, et le mode de démonstration est analogue au précédent.

Un raisonnement semblable au précédent montre aussi:

COROLLAIRE 5.5. Dans l'ensemble $C_+^\infty([0,1] \times R) \times C_\theta^\infty([0,1] \times R) \times R \times R \times R_+$, il existe un ouvert (non vide) constitué d'éléments (g,h,x_0,x_1,T) pour lesquels le problème paramétré associé n'a pas de solution.

De plus, en considérant également des "situations non génériques", il est intéressant de noter:

COROLLAIRE 5.6. Soit un couple critère $(g,h) \in E$. Une condition nécessaire et suffisante pour que (g,h) appartienne à E_s est que le champ hamiltonien associé ne présente pas de singularité de type Ic ou IIc_{22}.

Finalement, en construisant un chemin $({}_\tau g, {}_\tau h)_{\tau \in [0,1]} \subset C_+^\infty(R) \times C_\theta^\infty(R)$, on montre:

PROPOSITION 5.1. L'ensemble E_s est connexe.

COROLLAIRE 5.7. Etant donné deux couples critères $(\bar{g},\bar{h}) \in E_s$ et $(\tilde{g},\tilde{h}) \in E_s$ tels que le couple $(\bar{g},\tilde{g}) \in [C_+(R)]^2$ est générique, il existe un chemin $({}_\tau g, {}_\tau h)_{\tau \in [0,1]}$, d'extrémités $(\bar{g},\bar{h})$ et $(\tilde{g},\tilde{h})$, générique et inclus dans l'ensemble E_s.

Les hypothèses générales, la définition du problème (P), des ensembles Σ_i, et le théorème de relaxation sont valables pour toute valeur de $n \geq 1$; mais pour $n > 1$, l'équation H = constante ne suffit plus à caractériser les bitrajectoires, elle décrit simplement une hypersurface sur laquelle s'enroulent les bisolutions. Dans [9], les résultats du paragraphe 2 sont étendus au cas $n = 2$ pour des fonctions de type séparé. Puis on montre que l'ensemble S des points où le champs hamiltonien présente des types de singularités génériques est une union de sous-variétés de classe $C\infty$ et de dimension 2 dans R^4; d'où l'existence de bisolutions relaxées (et non originelles) dont les intersections respectives avec S sont des courbes (situation compètement différente du cas $n = 1$ dans lequel l'ensemble S est de dimension 0). Enfin, on détermine une classe de fonctions telles qu'il existe un triplet (x_0, x_1, T) pour lequel le problème (P) associé n'a pas de solution. Pour les fonctions restantes, l'étude est en cours.

BIBLIOGRAPHIE

1. Cerf, *Sur les difféomorphismes de la sphère:* $\Gamma_4 = 0$, Springer Lecture Notes 53.

2. Clarke, *Necessary conditions for a general control problem*, preprint, University of British Columbia.

3. Ekeland, Discontinuité de champs hamiltoniens et existence de solutions optimales en calcul des variations, *Publications Mathématiques de l'I.H.E.S.*, 1978.

4. Ekeland, *Cours de 3ème cycle de Mathématiques de la Décision*, Cahiers de Mathématiques de la Décision, Université Paris IX-Dauphine, 1977.

5. Golubitsky et Guillemin, Stable mappings and their singularities, *Graduate Texts in Mathematics 14:* (1973) 50-59.

6. Hirsch, Differential-topology, *Graduate Texts in Mathematics 33:* (1976).

7. Levine, Singularities of differentiable mappings, Liverpool Symposium on Singularities, *Springer Lecture Notes 192:* (1971) 1-89.

8. Mather, Stability of C^∞ mappings, V: Transversality, *Advances in Mathematics 4:* (1970) 301-336.

9. Ortmans, *Singularités de champs hamiltoniens, lien avec l'existence de solutions optimales en calcul des variations*, Cahiers de Mathématiques de la Décision, Université de Paris IX - Dauphine, 1978.

10. Pallu de la Barrière, *Cours d'automatique théorique*, Dunod, Paris, 1966.

NON-CONVEX PROGRAMMING

A NOTE ON NON-CONVEX OPTIMIZATION

H. Th. JONGEN
Universität Hamburg

In this paper we present some ideas concerning differentiable optimization problems in $\mathbb{R}^n$, using tools of global analysis.

1. Introduction.

Let $C^k(\mathbb{R}^n,\mathbb{R})$ denote the space of k-times continuously differentiable real-valued functions on the Euclidean $\mathbb{R}^n$, $\langle.,.\rangle$ being the standard inner product. We consider optimization problems of the following type:

$$\text{Minimize } f \quad \text{w.r.t. } M[h_i g_j],$$

where f, h_i, $g_j \in C^\infty(\mathbb{R}^n,\mathbb{R})$, $i \in I = \{1,\ldots,m\}$, $m < n$, $j \in J = \{1,\ldots,s\}$ and $M[h_i,g_j] = \{x \in \mathbb{R}^n | h_i(x) = 0,\ g_j(x) \geq 0,\ i \in I,\ j \in J\}$. The constraint-set $M = M[h_i,g_j]$ is called regular, if for every $\bar{x} \in M$ the set of gradients $\{Dh_i(\bar{x}), Dg_j(\bar{x}),\ i \in I,\ j \in J_0(\bar{x})\}$ is linearly independent $-J_0(\bar{x}) = \{j \in J | g_j(\bar{x}) = 0\}-$. From now on M will always be regular.

Let $\Sigma^k = \{x \in M | J_0(x) = n - m - k\}$. Then every nonempty Σ^k

is a smooth k-dim. submanifold (in the "open" sense) of $\mathbb{R}^n$ consisting of a countable number of connected components $\Sigma_1^k, \Sigma_2^k, \ldots$ called k-dim. strata. We may consider M as having been built up by all these Σ_i^k.

An $\bar{x} \in M$ is called a critical point for $f|_M$ if $\bar{x}$ is critical for the restriction of f to the stratum that contains $\bar{x}$ (cf. [4], [6]). In formulas: $\bar{x}$ is critical for $f|_M$ if

$$(1.1) \qquad Df(\bar{x}) = \sum_{i\in I} \lambda_i Dh_i(\bar{x}) + \sum_{j\in J_0(\bar{x})} \mu_j Dg_j(\bar{x}), \ldots$$

λ_i, μ_j being called Lagrange parameters.

If all $\mu_j \geq 0$ in (1.1), then we call $\bar{x}$ a (+) Kuhn-Tucker point.

Let A be an n×n-matrix viewed as a linear transformation and L a linear subspace of $\mathbb{R}^n$ with orthonormal basis $\{v_1, \ldots, v_d\}$. The determinant of A restricted to L will be the determinant of $(\alpha_{ij})_{i,j = 1,\ldots,d}$, where $\alpha_{ij} = \langle v_i, Av_j \rangle$. We define A to be nondegenerate on L if the determinant of A restricted to L is unequal to zero.

Taking (1.1) into account, a critical point $\bar{x}$ is called nondegenerate if:

$$(\text{ND1}) \qquad \mu_j \neq 0, \quad j \in J_0(\bar{x})$$

$$(\text{ND2}) \qquad D^2 f(\bar{x}) - \sum_{i\in I} \lambda_i D^2 h_i(\bar{x}) - \sum_{j\in J_0(\bar{x})} \mu_j D^2 g_j(\bar{x})$$

is nondegenerate on $\bigcap_{i\in I} \operatorname{Ker} Dh_i(\bar{x}) \cap \bigcap_{j\in J_0(\bar{x})} \operatorname{Ker} Dg_j(\bar{x})$,

where D^2 stands for the matrix of second derivatives and $\operatorname{Ker} v = \{w \in \mathbb{R}^n \mid \langle v,w \rangle = 0\}$.

For $\bar{x} \in M$, $|J_0(\bar{x})| = p$, M admits a local parametrization by means of local coordinates $(y_1, \ldots, y_n)$ having the property

$$(1.2)\qquad \begin{aligned} &y_k = \dots = y_m = 0 \\ &y_{m+1} \geq 0, \dots, y_{m+p} \geq 0 \end{aligned}$$

If $\bar{x}$ is a nondegenerate critical point for $f|_M$, $|J_0(\bar{x})| = p$, there exist local coordinates $(y_1,\dots,y_n)$ according to (1.2) such that f has the form:

$$(1.3)\qquad f(0,\dots,0,y_{m+1},\dots,y_n)$$

$$= f(\bar{x}) - \sum_{i_1=m+1}^{k} y_{i_1} + \sum_{i_2=k+1}^{m+p} y_{i_2} - \sum_{i_3=m+p+1}^{m+p+1} y_{i_3}^2 + \sum_{i_4=m+p+l+1}^{n} y_{i_4}^2 .$$

In (1.3) the number of negative (positive) linear terms is called the linear index LI (linear co-index LCI) and the number of negative (positive) quadratic terms is called the quadratic index QI (quadratic co-index QCI) of the critical point $\bar{x}$. The total index TI (total co-index TCI) is defined as to be LI + QI(LCI + QCI). The total index may be viewed as the maximal number of independent admissible directions of decrease for $f|_M$ at $\bar{x}$. The nondegenerate critical point $\bar{x}$ is a local minimum (maximum) for $f|_M$ iff TI = 0 (TCI = 0), and $\bar{x}$ is a (+) Kuh-Tucker point iff LI = 0.

We call $f|_M$ nondegenerate if all critical points for $f|_M$ are non-degenerate; if $f(\bar{x}) \neq f(\bar{y})$ for any two distinct critical points $\bar{x},\bar{y}$, then f is called separating. By the C^k-topology we denote the strong C^k (Whitney)--topology on $C^\infty(\mathbb{R}^n,\mathbb{R})$; a typical base-neighborhood of zero function is given by

$$\{\phi \in C^\infty(\mathbb{R}^n,\mathbb{R}) \mid \left|(\partial^{\alpha_1+\dots+\alpha_n}/\partial_{x_1}^{\alpha_1}\dots\partial_{x_n}^{\alpha_n})\phi(x)\right| < \varepsilon(x)$$

$$\text{for all } x \in \mathbb{R}^n \text{ and } \Sigma\alpha_i \leq k\},$$

where $\varepsilon(x) > 0$ and continuous. The C^k-topology for the product space $C^\infty(\mathbb{R}^n,\mathbb{R})^{1+m+s}$ is defined in an obvious way.

2. Genericity and Structural Stability

GENERICITY THEOREM. There exists a C^2-open and dense subset U of $C^\infty(\mathbb{R}^n,\mathbb{R})^{1+m+s}$ such that for $(f,h_i,g_j,\ i\in I,\ j\in J)\in U$ either $M[h_i,g_j]=\emptyset$ or $M=M[h_i,g_j]\neq\emptyset$, M is regular and $F|_M$ is nondegenerate. Furthermore, in the last case, f can be C^2-approximated arbitrary well by an $\tilde{f}$ such that $\tilde{f}|_M$ is nondegenerate and separating.

STRUCTURAL STABILITY THEOREM. Let $M=M[h_i,g_j]$ be regular, compact and $f|_M$ be nondegenerate, separating. Then there exists a C^2-neighborhood 0 of $(f,h_i,g_j,\ i\in I,\ j\in J)$ and for every $(\tilde{f},\tilde{h}_i,\tilde{g}_j,\ i\in I,\ j\in J)\in 0$ there exists C^∞-diffeomorphisms $\tilde{\phi}:\mathbb{R}^n\to\mathbb{R}^n$, $\tilde{\psi}:\mathbb{R}\to\mathbb{R}$ with the following properties:

(1) $\tilde{\phi},\tilde{\psi}$ differ from the identity only on a compact set,

(2) $\tilde{\phi}$ maps $M[h_i,g_j]$ onto $\tilde{M}=M[\tilde{h}_i,\tilde{g}_j]$,

(3) $\tilde{f}|_{\tilde{M}}=\tilde{\psi}\circ f\circ\tilde{\phi}^{-1}|_{\tilde{M}}$.

In [4] these theorems are proved by means of tools from differential topology. As an illustration of the Structural Stability Theorem in Figure 2.1, locally, some typical level curves of three functions of two variables are depicted. Note that in Figure 2.1.b the function is not separating.

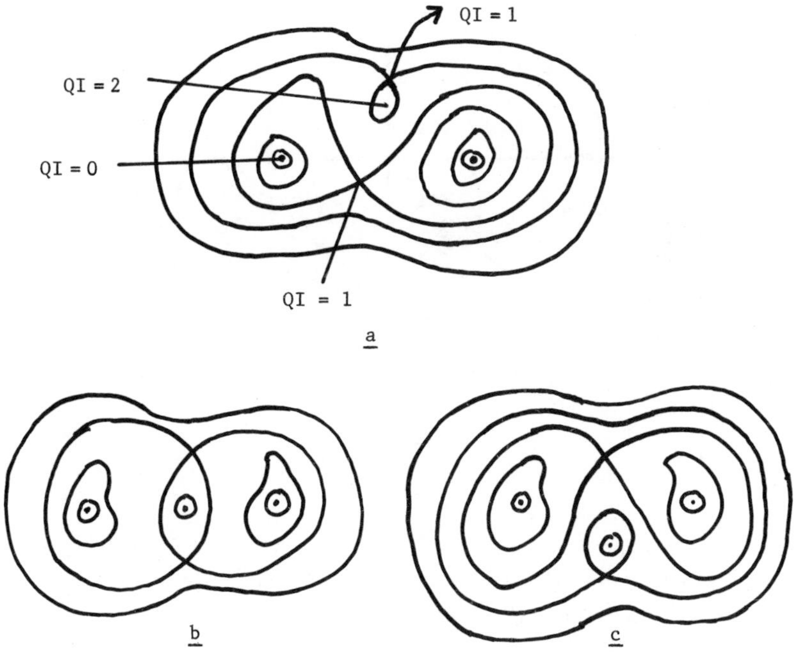

FIGURE 2.1

3. Decomposition

Let $M[h_i,g_j]$ be regular, compact and $f|_M$ be nondegenerate, separating. For $a \leq b$ we define $M_a^b = \{x \in M | a \leq f(x) \leq b\}$, $M^b = \{x \in M | \ f(x) \leq b\}$. Let us consider firstly some lower level sets of the function in Figure 2.1.a for increasing function values $c_1,\ldots,c_9$.

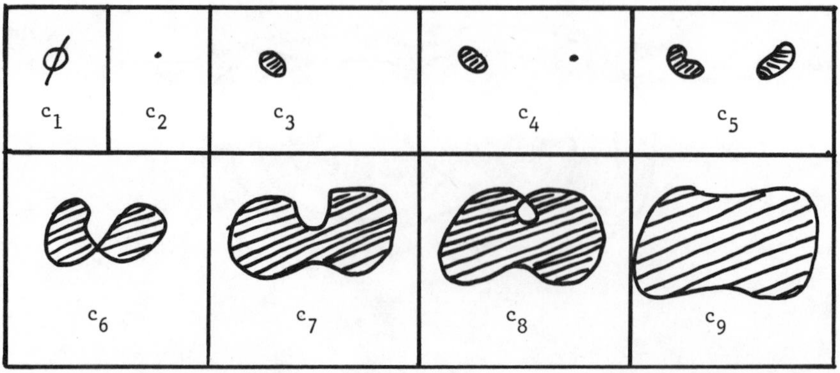

FIGURE 2.2

A typical situation in three dimensions is shown in Figure 2.3, where some level sets (suitably scaled) of a function of three variables are depicted for increasing function values $c_1,\dots,c_{10}$.

Interesting for the search for several local minima of $f|_M$ are those critical points at which the number of connected components of the lower level sets decreases for increasing function value; c.f.

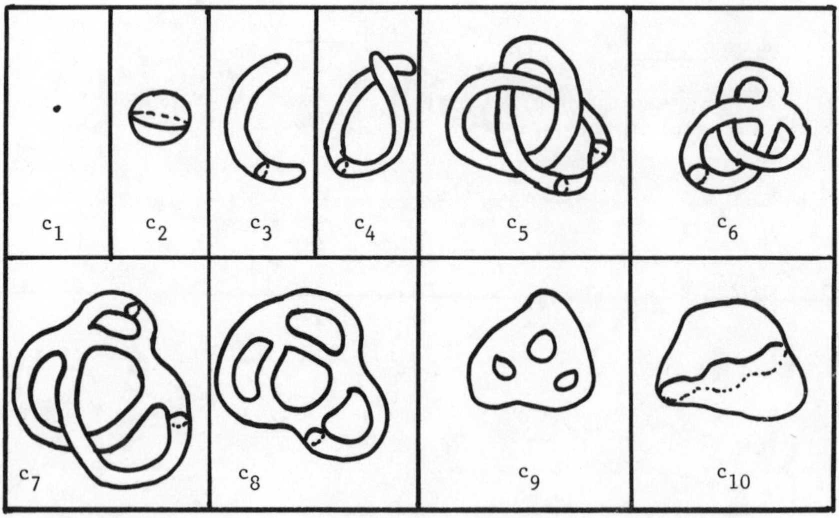

FIGURE 2.3

Figure 2.2, c_5, c_6. This leads to the following theorem, which can be proved by means of Morse Theory (cf. [1], [4], [6], [10]). Let $\$$ denote the number of connected components.

DECOMPOSITION THEOREM. Let $a < b$. If M_a^b contains no critical points, then $\$M^b = \$M^a$. Suppose that M_a^b contains exactly one critical point $\bar{x}$ with indices (LI,QI,LCI,QCI) and $a < f(\bar{x}) < b$. Then we have

$$\text{if } LI \neq 0: \quad \$M^b = \$M^a,$$

$$\text{if } LI = 0, \quad QI \neq 0,1: \quad \$M^b = \$M^a,$$

$$\text{if } LI = 0, \quad QI = 0: \quad \$M^b = \$M^a + 1 \quad \text{(local minimum)},$$

$$\text{if } LI = 0, \quad QI = 1: \quad \text{either} \quad \$M^b = \$M^a \quad \text{or} \quad \$M^b = \$M^a - 1.$$

4. The Graphs 0-1-0, 0-k-0

We consider a regular, compact, connected constraint set $M = M[h_i]$ (i.e. no inequality constraints are involved. Let $f|_M$ be nondegenerate, separating. Viewing M as a smooth manifold of dimension k, let $\mathcal{R}$ be a Riemannian metric on the tangent bundle of TM of M. Then $\mathcal{R}$ gives rise to a gradient vectorfield for f on TM, which induces a dynamical system Φ on M ("integral curves" of the gradient vectorfield). At each critical point of $f|_M$ we may define a stable and unstable manifold w.r.t. Φ (cf. [11]) and we suppose, without loss of generality, that all stable and unstable manifolds intersect transversally.

The graph 0-1-0 is defined as follows (cf. [5]): The vertex set corresponds to the set of critical points for $f|_M$ having quadratic index 0,1. In this way each vertex is labeled 0 or 1. Two vertices with the same label are not connected by an edge. Two vertices with different label are connected by an edge iff the corresponding critical points are connected (as limit points) by a trajectory of the dynamical system Φ.

The graph 0-k-0 is defined completely analogous. Note: A critical point having quadratic index k is a local maximum.

THEOREM. The graphs 0-1-0, 0-k-0 are connected.

For unconstrained problems, i.e. $M = \mathbb{R}^n$, this theorem holds if obvious appropriate additional conditions for f are taken into account. The search for critical points having quadratic index 1 could be done by means of a desingularized version of Newton's method (cf. [2], [3], [7], [12]). Then at least the critical points of f of odd quadratic index and those of even quadratic index will be separated from each other. In [9] a different approach is proposed by means of a vectorfield obtained by partial reflection of the gradient vectorfield of f.

5. A Homotopy Approach

We consider a regular, compact, constraint-set $M = M[h_i, g_j]$ and functions f_0, f_1 $C^\infty(\mathbb{R}^n, \mathbb{R})$, $f_i|_M$ nondegenerate, $i = 0,1$. Let $F \in C^\infty([0,1] \times \mathbb{R}^n, \mathbb{R})$, $F(i,\cdot) = f_i$, $i = 0,1$. We call F a smooth homotopy between f_0, f_1. The family of all smooth homotopies between f_0, f_1 will be denoted by $H(f_0, f_1)$, endowed with the induced C^3-topology of $C^\infty([0,1]$ $\mathbb{R}^n, \mathbb{R})$. Let $F \in H(f_0, f_1)$ and $(t,x) \in [0,1] \times M$. The pair (t,x) is called a (nondegenerate) critical point for $F|_M$ if x is a (nondegenerate) critical point for $F(t,\cdot)|_M$.

There exists a C^3-open and dense subset $H_r(f_0, f_1)$ of $H(f_0, f_1)$ such that the critical point set S of $F|_M$ for $F \in H_r(f_0, f_1)$ has a natural nice structure (cf. [8]). In particular, for $F \in H_r(f_0, f_1)$ there are exactly two types of degenerate critical points $(\bar{t}, \bar{x})$:

Type 1: exactly one Lagrange parameter μ_j vanishes $(j \in J_0(\bar{x}))$,

Type 2: a fold-singularity.

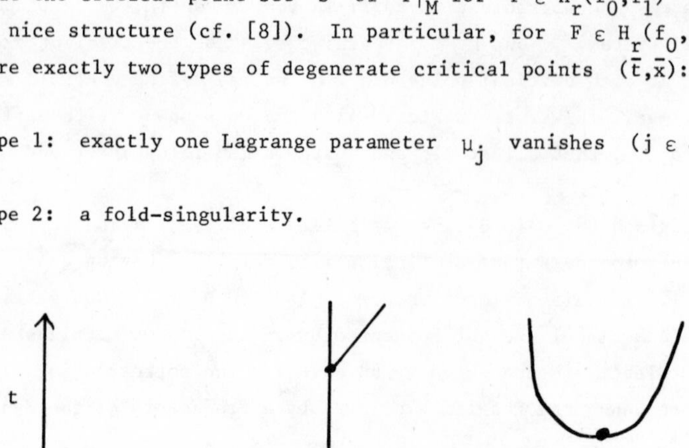

Type 1 Type 2

FIGURE 5.1

In [8] the critical point set of an $F \in H_r(f_0,f_1)$ is discussed in detail. Suppose that $(\bar{t},\bar{x})$ is a fold-singularity for $F|_M$ at which a pair of critical points with $TI = 0$, resp. $LI = 0$ and $QI = 1$ is created (or cancelled) in the increasing direction of t. Then at $(\bar{t},\bar{x})$ we have a direction of decrease of third order to compute, by means of a descending method, a local minimum for $F(\bar{t},\cdot)|_M$.

REFERENCES

1. D. Braess, Morse-theorie für berandete mannigfaltigkeiten, *Math. Ann. 208:* (1974) 133-148.

2. F. H. Branin, Jr., Widely convergent method for finding multiple solutions of simultaneous nonlinear equations, *IBM J. Res. Dev. 16:* (1972) 504-522.

3. L. C. W. Dixon and G. P. Szego (Eds), *Twoards global optimization*, North Holland Publishing Company (1975).

4. H. Th. Jongen, Dissertation, Twente University of Technology, The Netherlands (1977).

5. H. Th. Jongen, Zur Geometrie endlichdimensionaler nichtknovexer optimierungsaufgaben, *ISNM 36, Birkhäuser Verlag Basel:* (1977) 111-136.

6. H. Th. Jongen and F. Twilt, On decomposition and structural stability in non-convex optimization, *ISNM 46, Birkhäuser Verlag Basel:* (1979) 162-183.

7. H. Th. Jongen, P. Jonker and F. Twilt, On Newton flows in optimization. In: *Methods of Operations Research 31,* Eds. W. Oettli and F. Staffens, (1979), 345-359.

8. H. Th. Jongen, P. Jonker and F. Twilt, On deformation in optimization. To appear in *Proceedings of the IV. Symposium über Operations Research* (Saarbrücken, West Germany, 10-12 Sept. 1979).

9. H. Th. Jongen and J. Sprekels, The index k-stabilizing differential equation, to appear.

10. J. Milnor, *Morse theory*, Study 51, Princeton Univ. Press (1963).

11. S. Smale, On gradient dynamical systems, *Ann. of Math. 74:* (1961) 199-206.

12. S. Smale, A convergent process of price adjustment and global Newton methods, *J. Math. Econ. 3:* (1976) 107-120.

GAME THEORY

STABILITY BY THREATS AND COUNTERTHREATS IN NORMAL FORM GAMES

G. LAFFOND

H. MOULIN

C.E.R.E.M.A.D.E.
Université de Paris IX Dauphine

ABSTRACT

We define a stability concept for n-person games in normal form based on cooperative deterrence. Coalitions deter each other from deviating moves by mutual threats and counterthreats. Our concept involves neither side-payments nor interpersonal comparison of utility. It is proved to be non empty for all finite normal form games.

ACKNOWLEDGEMENTS

We are very indebted to R. Rosenthal and Y. Younes for various fruitful comments. We also wish to thank C. D'Aspremont and L. A. Gerard-Varet.

I. Introduction

In normal form games the different stability requirements leading to the various equilibrium concepts can be ordered among three levels with increasing sophistication: At the first level, the stability of a particular outcome means that no player or coalition of players has an incentive to deviate; At the second level it means that every player, or coalition of players, with an incentive to deviate faces a threat--that is an expected reaction of the nondeviating players--which deters him from deviating: Finally, at the thrid level, one assumes that any deviation from which a players, or coalition of players, cannot be deterred by a threat, is likely to raise up a strategic move-named couterthreat--of some other players. The stability requirements is that by foreseeing the counterthreat the deviating players are eventually deterred. At the first level we find a solution in the strict sense for two person games of (Luce and Raiffa [6]), extended to n person games as Aumann's strong equilibrium (see [1]), at the second level is the core of the α-characteristic function form (see [1]) and at the third level are the various versions of the von Neumann Morgenstern solution (see Luce and Raiffa [6], Vickrey [10], Harsanyi [5], and Roth [9], as well as the different bargaining sets (see Aumann and Maschler [2], Billera [3], d'Aspremont [4],...).

Clearly the first level stability requirements are so stringent that they usually cannot be met. On the contrary, stability by threats can normally be achieved in every two person normal game (every Pareto optimal individually rational outcome can be stabilized by a suitable pair of threats--this is elementary see e.g. Moulin [7]). It is not pathological that the α-core of a n person game is non empty but in some three player games it is empty. It is the case for instance in any normal form version of a simple majority game where a Condorcet effect arises. This justifies the use of the more subtle stability concepts of the third level.

As Rosenthal [8] pointed out, the spirit of the bargaining sets is to achieve stability by *retentiveness*: The counterthreat (called "counterobjection" in this context) is that each of the non deviating players can retain his original utility level. Another possible approach is to achieve stability by deterrence: The counterthreat is now that some of the non deviating players can break down the deviating coalition by forcing a loss upon at least one of its members (while possibly improving upon the utility of every other deviating player). These two approaches are different: In the bargaining set analysis, the counterobjection has no reason to break down the objection (that is the deviating coalition) whereas in the deterrence set analysis (Section 2 and Section 3 below) it may happen that in order to enforce stability some counterthreatening players have to suffer a voluntary loss. Thus the bargaining set stability requirement may fail to deter the players from deviating whereas in the punishment set analysis the players must sometimes heavily violate their own interest for the sake of collective stability.

Stability by deterrence is the key idea of Vickrey's self-policing imputation sets (see [10]) and was also studied by Rosenthal [8]. In contrast with the huge amount of results about the bargaining set, these papers contain no significant mathematical result (in particular no existence theory).

The aim of this paper is to display an *always non empty* stability concept which generalizes the usual imputation set of two players games. In contrast with most of the cooperative theory, we do not look for a small solution concept, e.g. a unique selection of an outcome like the nucleolus, but for a large subset of the imputation playing the role of an initial framework for the subsequence bargaining: This subset is large because the players enter a coalition only if this is worthwhile, no matter what the reaction of the rest of the world--thus even if the later "commits suicide" for the sake of stability.

This paper is organized as follows: In Section 2 we restrict ourselves to three person games, where analysis of threats and counterthreats is more simple. There we define a first stability concept: The *three-person deterrence* set and prove it is always non-empty.

We go to the general case in Section 3 where a *global deterrence set* is defined and proved again to be always non-empty.

II. The Three-Person Deterrence Set

A three-person normal form game is a 6 - uple $(X_1,X_2,X_3;u_1,u_2,u_3)$ where X_i, $i = 1,2,3$, is the strategy of player i and u_i, $i = 1,2,3$, his or her ordinal utility function defined on $X_1 \times X_2 \times X_3$.

An outcome of the game is any element $x = (x_1,x_2,x_3)$ of $X_1 \times X_2 \times X_3$. A coalition of players is a non empty subset S of $\{1,2,3\}$. Given a coalition S, a member i of S, and a joint strategy $x_S = (x_i)_{i \in S}$ of coalition S we denote by $\underline{u}_i(x_S)$ the secure utility level of player i when coalition S uses his joint strategy x_S:

$$\underline{u}_i(x_S) = \inf_{x_{\{1,2,3\}\setminus S}} u_i(x_S, x_{\{1,2,3\}\setminus S})$$

where $x_{\{1,2,3\}\setminus S}$ varies over the set of all joint strategy of the complement coalition of S.

An outcome x is said to be an *imputation* if it is together individually rational and Pareto optimal:

$$\forall\, i = 1,2,3: \sup_{y_i \varepsilon X_i} \underline{u}_i(y_i) \leq u_i(x) \quad \text{(individual rationality)}$$

$$\nexists\, y \quad X_1 \times X_2 \times X_3 \begin{cases} \forall\, i = 1,2,3: u_i(y) \geq u_i(x) \\ \qquad\qquad\qquad\qquad \text{(Pareto optimality)} \\ \exists\, i = 1,2,3: u_i(y) > u_i(x) \end{cases}$$

We denote by I the set of all imputations of the game. It is non empty when X_i, $i = 1,2,3$ are all finite.

Let $x \in I$ be an imputation. A threat against x is a pair (S, y_S) where S is a coalition, and y_S a joint strategy of coalition S such that:

$$\forall\, i \in S \quad \underline{u}_i(y_S) > u_i(x) .$$

Thus a threatening coalition can improve upon the utility of its members whatever is the reaction of the rest of the world. Since x is an imputation, a threatening coalition actually is a two-person coalition.

We denote by $O_{12}(x)$ the set of all threats $(\{1,2\}, y_{\{1,2\}})$

against x, and define similarly the notations $O_{23}(x)$ and $O_{13}(x)$.

Thus, the (possibly empty) α-core of the game is the set of those imputations x such that $O_{ij}(x)$ is empty for all two-person coalitions.

DEFINITION. The three-person deterrence set of game $(X_1,X_2,X_3,u_1,u_2,u_3)$ is the set of those imputations $x \in I$ such that the following properties hold true:

i) To every threat $(\{1,2\},y_{\{1,2\}}) \in O_{12}(x)$ by coalition $\{1,2\}$ against x there exists a *counterthreat* (z_1,z_3) by coalition $\{1,3\}$ and/or a counterthreat (z_2,z_3) by $\{2,3\}$.

A counterthreat (z_2,z_3) by $\{2,3\}$ against $(\{1,2\},y_{\{1,2\}})$ is a joint strategy of coalition $\{2,3\}$ such that

(1) $$\underline{u}_2(z_2,z_3) = \inf_{z_1 \in X_1} u_2(z_1,z_2,z_3) > \inf_{y_3 \in X_3} u_2(y_1,y_2,y_3) = \underline{u}_2(y_1,y_2)$$

and

(2) $$\sup_{z_1 \in X_1} u_1(z_1,z_2,z_3) < u_1(x)$$

and a similar definition holds for a counterthreat (z_1,z_3) by $\{1,3\}$ against $(\{1,2\},y_{\{1,2\}})$.

ii) To every threat $(\{2,3\},y_{\{2,3\}}) \in O_{23}(x)$ by $\{2,3\}$ against x there exists a counterthreat (z_1,z_2) by $\{1,2\}$ and/or a counterthreat (z_1,z_2) by $\{1,3\}$.

iii) To every threat in $O_{13}(x)$, there exists a counterthreat by $\{1,2\}$ and/or a counterthreat by $\{2,3\}$.

Let us comment on the stability properties described by the above definition. First of all if x belongs to the α-core of the game no threat exists against it so x belongs to the three-person deterrence set. The interesting case is when x is an imputation but does not belong to the α-core so that some two players coalition, say $\{1,2\}$, has a threat $y_{\{1,2\}}$ against x. What is then a counterthreat (z_2,z_3) by $\{2,3\}$ against $(\{1,2\},y_{\{1,2\}})$? It is a move by coalition $\{2,3\}$ intended to break down the threatening coalition $\{1,2\}$ by bribing player 2, while deterring player 1. Namely in order to split coalition $\{1,2\}$ player 3 tells to player 2: If you persist in the threat (y_1,y_2) I can force four utility level as low as $\underline{u}_2(y_1,y_2)$; if, on the other hand you form the counterthreat (z_2,z_3) with me, your utility level cannot go below $\underline{u}_2(z_2,z_3)$. Since $\underline{u}_2(z_2,z_3) > \underline{u}_2(y_1,y_2)$ (inequality (1)) you have an incentive to join me: This is how player 2 is bribed.

Next inequality (2) means that whenever the counterthreat (z_2,z_3) is formed, player I will never recover his or her original utility level $u_1(x)$: Therefore player 1, anticipating the counterthreat (z_2,z_3), is deterred from entering the threat (y_1,y_2).

Remark that the only incentive for player 3 to enter the counterthreat (z_2,z_3) is the deterring purpose itself: If player 3 does not react, his utility will be, in any case, lower than it was in the initial imputation.

EXAMPLE 1. It is a normal form majority game:

$$X_1 = \{2,3\}, \quad X_2 = \{3,1\}, \quad X_3 = \{1,2\},$$

$$(u_1,u_2,u_3)(2,1,x_3) = (10,5,0) \qquad \forall\, x_3 \in X_3$$

$$(u_1,u_2,u_3)(x_1,3,2) = (0,10,5) \qquad \forall\, x_1 \in X_1$$

$$(u_1,u_2,u_3)(3,x_2,1) = (5,0,10) \qquad \forall\, x_2 \in X_2$$

$$(u_1,u_2,u_3) \qquad = (0,0,0) \qquad \text{otherwise.}$$

This game has six imputations and an empty core.

The three-person deterrence set is equal to the imputation set. To prove that for instance the outcome $(2,1,x_3)$ is stable it suffices to remark that to the threat $(\{2,3\},(3,2))$, there is the counterthreat $(3,1)$ by coalition $\{1,3\}$.

THEOREM 1. Every three-person normal form game with finite strategy sets has a non empty three-person deterrence set.

Proof. Let X_i, $i = 1,2,3$ be three finite sets and let $(X_1,X_2,X_3,u_1,u_2,u_3)$ be a game with an empty three-person deterrence set. We will derive a contradiction by constructing a sequence $x^1,\ldots,x^t,\ldots$ within the imputation set I such that $x^t \neq x^{t'}$ for every $t \neq t'$.

We shall say that the threat (s,y_S) against x is maximal if there is no joint strategy z_S of coalition S such that:

$$(3) \qquad \forall\, i \in S \quad \underline{u}_i(z_S) > \underline{u}_i(y_S).$$

We let the reader check that if the threat (S,y_S) against x has no counterthreat, then every pair (S,z_S) verifying (3) is a threat against x with no counterthreat. By assumption against any imputation x^o there exists a maximal threat (S^o,y_{S^o}) with no possible counterthreat. Clearly S^o is a two-person coalition. Given y_{S^o}, there exists an imputation x^1 such that:

$$\forall\, i \in S^o \quad u_i(x^1) \geq \underline{u}_i(y_{S^o})$$

We construct now by induction a sequence s^o, (S^o,y_{S^o}), x^1, (S^1,y_{S^1}), $x^2,\ldots,(s^{t-1},y_{S^{t-1}}),x^t,\ldots$ such that x^k is an imputation and (S^k,y_{S^k}) a maximal threat against x^k with no counterthreat such that

$$\forall\, i \in S^k \quad u_i(x^k) \geq \underline{u}_i(y_{S^{k-1}}),$$

Suppose this sequence is constructed until x^t $(t \geq 1)$. We choose a maximal threat (S^t, y_{S^t}) against x^t with no possible counterthreat. By induction assumption $(S^{t-1}, y_{S^{t-1}})$ is a maximal threat against x^{t-1}. Therefore $S^t \neq S^{t-1}$: hence $S^{t-1} \setminus S^t$ contains exactly one player denoted i_t, and $S^{t-1} \cap S^t$ also contains a single player denoted j_t. Because y_{S^t} is a threat against x^t we have:

$$\underline{u}_{j_s}(y_{S^t}) > u_{j_t}(x^t) \geq \underline{u}_{j_t}(y_{S^{t-1}}).$$

Because there is no counterthreat to the threat $(S^{t-1}, y_{S^{t-1}})$ against x^{t-1} condition (2) of the definition yields:

$$\exists\, z_{u_t} : u_{i_t}(z_{i_t}, y_{S^t}) \geq u_{i_t}(x^{t-1}).$$

We select now x^{t+1} as an imputation dominating (z_{i_t}, y_{S^t}). Thus by construction our sequence $x^o, x^1, \ldots, x^t, \ldots$ verifies:

(4) $\quad \forall\, t \geq 1 \quad \forall\, i \in S^{t-1} \quad u_i(x^t) > u_i(x^{t-1})$

(5) $\quad \forall\, t \geq 1,$ denote $\{i_t\} = S^{t-1} \setminus S^t = (S^t)^c,$

then $u_{i_t}(x^{t+1}) \geq u_{i_t}(x^{t-1}).$

LEMMA. Fix t_0, t with t_0 t. Then

(6) $\quad \# \{i \in \{1,2,3\} \,/\, u_i(x^t) > u_i(x^{t_0})\} = 2.$

This lemma implies clearly that $x^t \neq x^{t_0}$ for every $t_0 < t$ and thus contradicts the finiteness of $X_1 \times X_2 \times X_3$. In order to prove the Lemma, we fix t_0 and consider condition (6) as property P_t. In view of (4) P_{t_0+1} holds true. We prove by induction that P_t holds true for every $t > t_0$. Suppose that $P_{t_0+1},\ldots,P_t$ hold true but P_{t+1} fails to be true. Because x^{t+1} and x^{t_0} are both imputations, there is at least one player i^o such that $u_i(x^{t+1}) \leq u_i(x^{t_0})$. Because P_{t+1} fails at least 2 players we verify this inequality. For simplicity we assume:

$$(7) \qquad u_1(x^{t+1}) \leq u_1(x^{t_0}); \qquad u_3(x^{t+1}) \leq u_3(x^{t_0})$$

By (4) the 2 players of S^t verify $u_i(x^t) < u_i(x^{t+1})$. In view of (7) these 2 players cannot be 1 and 3 (otherwise we would contradict P_t). We will then have for instance

$$u_1(x^t) < u_1(x^{t+1}); \qquad u_2(x^t) < u_2(x^{t+1}).$$

Therefore $u_1(x^t) \leq u_1(x^{t_0})$ and by P_t this implies:

$$u_2(x^{t_0}) < u_x(x^t); \qquad u_3(x^{t_0}) < u_3(x^t).$$

From the above inequalities we deduce $u_1(x^t) < u_1(x^{t_0})$ (otherwise x^{t_0} is not Pareto optimal) and we retain:

$$u_3(x^{t+1}) \leq u_3(x^{t_0})$$

$$(8) \qquad u_1(x^t) < u_1(x^{t_0})$$

$$S^t = \{1,2\}.$$

Because $S^t = \{1,2\}$ we have $\{i_t\} = (S^t)^c = \{3\}$.

By condition (5) and (8) we get:

$$(9) \qquad u_3(x^{t-1}) \leq u_3(x^{t+1}) \leq u_3(x^{t_0}) .$$

Because P_{t-1} holds true this implies

$$u_2(x^{t_0}) < u_2(x^{t-1}); \qquad u_1(x^{t_0}) < u_1(x^{t-1}).$$

Therefore $u_3(x^{t-1}) < u_3(x^{t_0})$ (otherwise x^{t_0} is not Pareto optimal). Moreover in view of (8) we obtain

$$u_1(x^t) < u_1(x^{t-1})$$

by condition (6) this yields $S^{t-1} = \{2,3\}$. We retain:

$$u_3(x^{t-1}) < u_3(x^{t_0})$$

$$u_1(x^t) < u_1(x^{t_0})$$

$$S^{t-1} = \{2,3\}.$$

From now on, the argument can be repeated: $i_{t-1} = 1$ yields $u_1(x^{t-2}) \leq u_1(x^t) < u_1(x^{t_0})$ and then (by P_{t-2}) $u_3(x^{t_0}) < u_3(x^{t-2})$ thus by (9) $S^{t-2} = \{1,2\}$:

$$u_3(x^{t-1}) < u_3(x^{t_0})$$

$$u_1(x^{t-1}) < u_1(x^{t_0})$$

$$S^{t-2} = \{1,2\}.$$

Insofar as $t - 2p + 1 > t_0$ we get $u_3(x^{t-2+1}) < u_3(x^{t_0})$ and $S^{t-2p+1} = \{2,3\}$. Similarly if $t - 2p > t_0$ we get $u_1(x^{t-2p}) < u_1(x^{t_0})$ and $S^{t-2p} = \{1,2\}$.

Suppose for instance $t - 2\bar{p} + 1 = t_0 + 1$. Then $i_{t_0+1} = 1$ and condition (5) yields: $u_1(x^{t_0}) \leq u_1(x^{t_0+2}) = u_1(x^{t-2(\bar{p}-1)}) < u_1(x^{t_0})$ a contradiction. Similarly if $t - 2\bar{p} = t_0 + 1$ then $i_{t_0+1} = 3$ and condition (5) yields: $u_3(x^{t_0}) \leq u_3(x^{t_0+2}) = u_3(x^{t-2\bar{p}+1}) < u_3(t_0)$ a contradiction again.

III. The Global Deterrence Set

The major drawback of the three-person deterrence set is that it cannot be extended as such to n person games unless we allow for the possibility that it becomes empty.

A successful generalization will be achieved by a more precise definition of a threat.

Let us introduce some definitions and notations:

An n-person normal form game is an object $\Gamma = (N,(X_i)_{i \in N},(u_i)_{i \in N})$ where N is the finite set of players and for all $i \in N$, X_i is the set of strategies of player i, and u_i, mapping $\prod_{i \in N} X_i = X^N$ into $\mathbb{R}$ measures the utility level of player i.

We call pay-off any vector $a = (a_i)_{i \in N}$ in $\mathbb{R}^N$ such that there exists $x \in X^N$ verifying $u_i(x) \geq a_i \; \forall \, i \in N$. Then we say that a_i is the pay-off to player i.

Let $C(N)$ be the set of non empty coalitions of players. For any coalition S we denote by $X^S = \prod_{i \in S} X_i$ the set of joint strategies of coalition S. For any two disjoint coalitions S and T and for any $z \in X^S$, $z' \in X^T$, we denote by (z,z') the element of $X^{S \cup T}$ such that $(z,z')_i = z_i$ if $i \in S$ and $(z,z')_i = z'_i$ if $i \in T$.

A coalition can threaten the *pay-off* a if there exists a joint strategy $x_S \in X^S$ that improves a^S whatever is the reaction of the complement coalition S^c. In order to formalize this idea we denote for all $S \in C(N)$ and all $i \in S$

$$S^c = \{i \in N / i \notin S\} ,$$

(10) $$\underline{u}_i(x_S) = \inf\{u_i(x_S, y),\ y \in S^{S^c}\} \quad \text{if} \quad S^c \neq \emptyset$$

$$\underline{u}_i(x_S) = u_i(x_S) \quad \text{if} \quad S^c = \emptyset .$$

We shall say that $b^S = (b_i^S)_{i \in S}$, $b^S \in \mathbb{R}^S$ is a guaranteed pay-off of coalition S if there exists $x_S \in X^S$ such that

(11) $$\underline{u}_i(X_S) \geq b_i^S \quad \text{all} \quad i \in S.$$

We denote by $A(S)$ the set of guaranteed pay-off of coalition S.

We will say that the triple (S, x_S, b^S) where $S \in C(N)$, $b^S \in A(S)$ (i.e. (11) holds true) is a *threat* of coalition S against pay-off a if:

(12) $$b_i^S > a_i \quad \text{all} \quad i \in S.$$

The α-core of game Γ is the set of pay-off against which there is no threat.

Notice. That in inequality (11) equality needs not to be true: This is so because b_i^S is interpreted as the utility level that player i is promised off when entering coalition S. Lowering b_i^S (still verifying (12)) will make it less easy for counterthreateners to break down coalition S (see Definition below).

Let us now describe counterthreats. We shall say that a triple

(T,z_T,C^T) is a *counterthreat* to threat (S,x_S,b^S) if:

$$T \in C(N), \quad T^c \cap S \neq \emptyset, \quad z_T \in X^T \text{ and } c^T \in A(T)$$

$$\text{(13)} \qquad \underline{u}_i(z_T) \geq c_i^T \quad \forall\, i \in T$$

$$c_i^T > b_i^S \quad \forall\, i \in T \cap S$$

Just as in the definition in Section 2, a counterthreat allows the members of $T \setminus S$ to bribe the payers in $T \cap S$ since they are promised a utility level c_i^T when joining the counterthreat, strictly better than their promised utility level b_i^S when keeping to the threat.

To the counterthreat (T,z_T,c^T) the payers of $S \cap T^c$ will react by joining some players of $(T \cup S)^c$ and form a new coalition W. If there exists a joint strategy z_W of the players in W guaranteeing a pay-off d^W such that

$$\text{(14)} \qquad \begin{aligned} &\underline{u}_i(z_T,z_W) \geq d^W \geq b^S \quad \forall\, i \in S \cap W = T^c \cap S \\ &\underline{u}_i(z_T,z_W) > a_i \quad \forall\, i \in W \cap S^c \end{aligned}$$

then coalition $S \cap T^c$ has got a successful reply to counterthreat (T,z_T,c^T) that eventually failed to deter the threat $(S,x_S b^S)$.

We denote by $R(a,(S,x_S,b^S),(T,z_T,c^T))$ the set of pairs (W,z_W) satisfying (14). If $R(a,(S,x_S,b^S),(T,z_T,c^T))$ is the empty set we shall say that the counterthreat (T,z_T,c^T) successfully *deters* the threat (S,x_S,b^S) against a.

DEFINITION. The *global deterrence set* of game Γ is the set of these pay-offs $a \in \mathbb{R}^N$ such that to every threat (S,x_S,b^S) against a one can associate at least one deterring counterthreat.

THEOREM 2. Suppose X_i, the strategy set of player i is finite for all $i \in N$. Then the global deterrence set of game Γ is non empty.

In two-person games, the set of imputations is never empty (when the strategy sets are finite). Therefore it is an a priori upper bound to all potential bargaining to select one specific outcome. For n-person games the global deterrence set, as intricated as its definition is, is a candidate for providing an a priori bound to negotiation among cooperative agents. The main subtleness in its definition is that a threatening coalition must promise to its members a certain utility level and must then balance two contradictory criteria: Improving b_S makes it more difficult for the rest of the world to bribe some members of S (see (13)) whereas lowering b_S makes it more difficult to deter the non bribed members of S (see (14)).

PROOF OF THEOREM 2. Let Γ be a n-person game where X_i is finite for all $i \in N$ and $D(\Gamma) = \emptyset$. If $x \in X^N$ we denote by $M(x)$ the set of those threats against pay-off $(u_i(x))_{i \in N}$ for which no deterring counterthreat exists. Fix now an element $\overset{o}{x} \in X^N$. We will first construct a sequence $(\overset{n}{x},(S_n,x_{S_n},\overset{n}{a}))n \in \mathbb{N}\setminus\{0\}$ for all n such that:

$$
(15) \qquad
\begin{aligned}
&(S_n,x_{S_n},\overset{n}{a}) \in M(\overset{n-1}{x}) \\
&(u_i(\overset{n}{x}) \geq \underline{u}_i(x_{S_n}) \quad \forall\, i \in S_n \\
&u_i(\overset{n}{x}) \geq \overset{p}{a}_i \quad \forall\, p \in \{1,\dots,n\}, \quad \forall\, i \in S_p\,.
\end{aligned}
$$

Fix $(S_1,x_{S_1},\overset{1}{a}) \in M(\overset{o}{x})$ and $t_1 \in X^{S^c_1}$, $\overset{1}{x} = (x_{S_1},t_1)$. One checks that $(\overset{1}{x},(s_1,x_{S_1},\overset{1}{a}))$ satisfies (1) for $n = 1$. Suppose that we have constructed $(\overset{n}{x},(S_n,x_{S_n},\overset{n}{a}))$ satisfying (15) for $1 \leq n \leq m$. Then take $(S_{m+1},x_{S_{m+1}},\overset{m+1}{a}) \in M(\overset{n}{x})$. We show that there exists $\overset{m+1}{x} \in X^N$

such that $(\overset{m+1}{x},(S_{m+1},x_{S_{m+1}},\overset{m+1}{a}))$ satisfies (15). Let us set:

$T_{m+1} = S_{m+1}, y_{m+1} = x_{S_{m+1}}$. We have

$$\underline{u}_i(y_{m+1}) \geq \overset{p}{a}_i \quad \forall\ i \in S_p, \quad p = m+1$$

$$\underline{u}_i(y_{m+1}) > \overset{p}{a}_i \quad \forall\ i \in S_p \cap T_{m+1}, \quad 1 \leq p \leq m$$

$$S_{m+1} \subset T_{m+1} .$$

We now prove the existence of two sequences $(T_k)_{m+1 \geq k \geq 1}$, $(y_k)_{m+1 \geq k \geq 1}$ such that, for all $k \in \{1,\dots,m+1\}$, we have:

$$\begin{aligned} & T_{h+1} \subset T_h \\ & T_h \supset S_{m+1} \cup \dots \cup S_h \\ (16) \qquad & Y_h \in X^{T_h} \\ & \underline{u}_i(y_h) \geq \overset{p}{a}_i \quad \forall\ p \in \{h, h+1, \dots, m+1\} \\ & \underline{u}_i(y_h) > \overset{p}{a}_i \quad \forall\ p \in \{1,\dots,h-1\} \quad \forall\ i \in S_p \cap T_h . \end{aligned}$$

Suppose that for some $k > 2$, $T_{m+1},\dots,T_k$ and $y_{m+1},\dots,y_k$, are constructed. We have

$$\underline{u}_i(y_k) > \overset{k-1}{a}_i \quad \forall\ i \in S_{k-1} \cap T_k .$$

So that $(T_k, y_k, (\underline{u}_i(y_k))_{i \in T_k})$ is a counterthreat to the threat $(S_{k-1}, x_{S_{k-1}}, \overset{k-1}{a})$. Then there exists $W \subset T_k^c$ and $z_W \in X^W$ such that:

$$\underline{u}_i(y_k,z_W) \geq a_i^{k-1} \quad \forall\, i \in W \cap S_{k-1} = S_{k-1} \cap T_k^c$$

$$\underline{u}_i(y_k,z_W) > u_k(x^{k-2}) \quad \forall\, i \in W \cap S_{k-1}^c .$$

Choosing $T_{k-1} = T_k \cup W$ and $y_{k-1} = (y_k,z_W)$, we have

$$\underline{u}_i(y_{k-1}) \geq \underline{u}_i(y_k) > a_i^p \quad \forall\, p \in \{1,\ldots,k-2\}, \quad \forall\, i \in T_k \cap S_p$$

$$\underline{u}_i(y_{k-1}) > u_i(x^{k-2}) \geq a_i^p \quad \forall\, p \in \{1,\ldots,k-2\}, \quad \forall\, i \in W \cap S_p ,$$

therefore

$$\underline{u}_i(y_{k-1}) > a_i^p \quad \forall\, p \in \{1,\ldots,k-2\}, \quad \forall\, i \in T_{k-1} \cap S_p ,$$

so that we have finally

$$T_{k-1} = T_k \cup W \supset T_k$$

$$T_{k-1} \supset S_{k-1} \cup S_k \cup \ldots \cup S_{m+1}$$

$$y_{k-1} \in X^{T_{k-1}}$$

$$\underline{u}_i(y_{k-1}) \geq \underline{u}_i(y_k) > a_i^p \quad \forall\, p \in \{k,\ldots,m+1\}, \quad \forall\, i \in S_p$$

$$\underline{u}_i(y_{k-1}) \geq a_i^{k-1} \quad \forall\, i \in S_{p-1}$$

$$\underline{u}_i(y_{k-1}) > a_i^p \quad \forall\, p \in \{1,\ldots,k-2\}, \quad \forall\, i \in S_p \cap T_{k-1} .$$

Implying that (16) holds for $h = k-1$.

Picking now y_{T_1} and $z \in X^{T_1^c}$, we set $x^{m+1} = (y_{T_1},z)$. This yields:

$$u_i(x^{m+1}) \geq u_i(x_{S_{m+1}}) \qquad \forall\, i \in S_{m+1}$$

$$u_i(x^{m+1}) \geq a_i^p \qquad \forall\, p \in \{1,\dots,m+1\}, \quad \forall\, i \in S_p .$$

Hence $(x^{m+1}, (S_{m+1}, x_{S_{m+1}}, a^{m+1}))$ satisfies (15) thus providing the existence of a sequence verifying (15).

One checks now that for all $n \in N$, $m \in N$, $n > M$, we have $x^m \neq x^n$. Namely

$$u_i(x^n) \geq a_i^{m+1} > u_i(x^m) \qquad \forall\, i \in S_{m+1} .$$

Since $x^n \in X^N$ $\forall\, n \in N$, we deduced that X^N contains infinitely many elements, a contradiction.

REFERENCES

1. R. Aumann, *A survey of cooperative games without side-payments.* Essays in Mathematical Economics in Honor of O. Morgenstern. Princeton University Press, 1967.

2. R. Aumann and M. Maschler, *The bargaining set for cooperative games. Advances in game theory.* Princeton University Press, 1964.

3. L. J. Billera, Existence of general bargaining sets for cooperative games without side-payments, *Bull. Amer. Math. Soc. 76:* (1970) 375-379.

4. C. D'Aspremont, *The bargaining set concept for cooperative games without side-payments.* Ph.D. Dissertation, Stanford University, 1973.

5. J. Harsanyi, An equilibrium-point interpretation of stable sets and a proposed alternative definition. *Management Science 20(11):* (1974) 1472-1495.

6. R. Luce and H. Raiffa, *Games and decisions,* Wiley and Son, 1957.

7. H. Moulin, Deterrence and cooperation: A classification of two-person games. To appear in the *European Economic Review* (1980).

8. R. Rosenthal, Cooperative games in effectiveness form. *Journal of Economic Theory 5(1):* (1972).

9. A. Roth, Subsolutions and the supercore of cooperative games. *Mathematics for Operation Research 1(1):* (1976).

10. W. S. Vickrey, Self-policing properties of certain imputation sets. *Annals of Mathematical Study 40:* (1959) 213-246.